Implementing IT in Construction

T0173940

Contrary to stereotype, the construction industry has embraced IT with some vigour. Computers are used effectively right across the sector and this use is increasing. A range of new issues have emerged in consequence.

Implementing IT in Construction is a practical book which draws on direct industrial experience and examines the role of IT within a range of enterprises operating in the construction and property industries. Emphasis is given to the human side of IT and the effects of the implementation of IT systems on those who use them. The functionality of the IT systems is considered, as is the design brief and the operation of the applications. Case reviews of a range of applications are discussed and issues arising from their implementation are explored. The benefits and pitfalls of implementing IT in the construction industry are surveyed and presented systematically so the reader can consider them in the light of their own experience.

Readers should get a good sense of the value of IT implementation and appreciate the benefits of a number of IT applications. Key drivers are presented for advanced students and for professionals who may have to face these issues in the future. This book is an essential guide which will aid the user in implementing their IT systems more effectively and efficiently.

James Sommerville holds a Chair in Construction Management within the School of the Built and Natural Environment at Glasgow Caledonian University. He is a chartered surveyor and builder with considerable industrial experience both in the UK and the Middle East. Throughout his career he has been immersed in the identification, development and implementation of bespoke IT systems for organisations large and small.

Nigel Craig is a Research Fellow within the School of the Built and Natural Environment at Glasgow Caledonian University. He gained extensive site experience within the construction industry prior to undertaking his post-graduate studies. He continues his research into IT in the industry and to secure his PhD.

Implementing IT in Construction

James Sommerville and Nigel Craig

 Routledge
Taylor & Francis Group
LONDON AND NEW YORK

First published 2006 by Taylor & Francis
This edition published 32013 by Routledge
2 Park Square, Milton Park, Abingdon, Oxon OX14 4RN

Simultaneously published in the USA and Canada
by Routledge
711 Third Avenue, New York, NY 10017

*Routledge is an imprint of the Taylor & Francis Group
an informa business*

© 2006 James Sommerville and Nigel Craig

Typeset in Times by
Integra Software Services Pvt. Ltd, Pondicherry, India

All rights reserved. No part of this book may be reprinted or reproduced or
utilised in any form or by any electronic, mechanical, or other means, now
known or hereafter invented, including photocopying and recording, or in
any information storage or retrieval system, without permission in writing
from the publishers.

British Library Cataloguing in Publication Data
A catalogue record for this book is available from the British Library

Library of Congress Cataloging in Publication Data
A catalog record has been requested for this book

ISBN10: 0-415-37051-5 (hbk)
ISBN10: 0-415-37052-3 (pbk)
ISBN10: 0-203-03039-7 (ebk)

ISBN13: 978-0-415-37051-6 (hbk)
ISBN13: 978-0-415-37052-3 (pbk)
ISBN13: 978-0-203-03039-4 (ebk)

CONTENTS

CHAPTER 4
Contemporary business processes: how IT improves these processes

CHAPTER 5
Capturing knowledge within the construction industry

CHAPTER 6
Capturing information at the project level: the reliance on paper

CHAPTER 7
IT tools: Electronic Document Management Systems (EDMS), extranets and mobile communication devices

CHAPTER 8
The implementation of IT within construction organisations

CHAPTER 9
Implementing IT: reviews of industry projects and applications

LIST OF FIGURES

LIST OF TABLES

PREFACE

There have been a great number of books written about the underpinning theories and approaches surrounding the use of IT. Some of these have included specific sections on the construction industry within them; others have alluded to the inclusion of the industry as part and parcel of the much broader industrial base operating within the UK.

We felt that there was a clear need for a book that looked specifically at IT implementation within the construction industry and provided useful commentaries on do's and don'ts along with lessons learnt when reviewing a number of cases where IT had been implemented. Other books have discussed the hardware and software available for use but failed to take this the 'extra mile' and show IT being put into practice in any scale. The implementation of IT of course demands that a range of issues surrounding, and affecting, the decision to go ahead with the implementation of the IT should be discussed.

The purpose of the book can best be described by saying that we, through real life experiences, have witnessed a range of construction organisations struggle to get to grips with business problems that could be resolved through appropriate use of IT, if only the companies had had something that could have guided them (even partially). Much of what was available was written very much from an 'IT specialist's' perspective. People operating within the majority of construction businesses are not IT specialists, so the materials available could at best be seen as being only partially helpful to them. This book sets out to redress this imbalance.

The materials within the chapters are intentionally set out to allow the widest possible level of engagement with them. Hopefully the further education and undergraduate cohorts will see the text as expanding and explaining many of the issues they are striving to grasp; also, we hope that the more advanced reader will see the case reviews of the implementations as providing valuable insights into the complexities found within a range of projects and organisations. For the industrial practitioners we hope that the open and frank discussions on each of the cases will offer them some guidance as to how best to work through their own problem areas or forthcoming IT agendas.

The question always arises as to what the book actually tells the reader; in answer to this it is the intention that the each chapter discusses and elaborates on a number of factors to be considered when driving forward decisions surrounding whether or not to implement IT. Also, the analysis of the cases enables us to shed light on many of the pro's and cons of IT implementation and compile a wide ranging list of Do's and Don'ts. The types of projects and cases where IT has been implemented are discussed and the approach used in each case is considered.

The book contains 10 Chapters; each of which adds to the general flavour of the discussion on IT implementation: Chapter 1 sets the broad landscape of IT and how it relates to the way construction is delivered; Chapter 2 considers the operation of the industry and the range of organisations likely to engage in IT activities; Chapter 3 focuses on the role humans play in engaging with the IT initiative and its effects on the humans; Chapter 4 concentrates on the processes executed on a day to day basis and how they may be improved through appropriate IT; Chapter 5 delves into the knowledge repositories found within the industry and how we manage such knowledge bases; Chapter 6 takes the issue of knowledge much further and looks at how we capture and process information at a range of organisational levels; Chapter 7 reviews arrange of applications (tools) found within the industry; Chapter 8 looks at IT implementation and the mistakes often made; Chapter 9 reviews the cases in some detail and sets out the approach adopted in each and; Chapter 10 provides a consolidation of what has been gleaned from the case reviews and provides suggestions as to what to or not to do.

The book is not intended to be the panacea to all IT issues. We openly admit to not being IT experts, this we would argue is our saving grace. Further, we have been limited by the need for brevity in many areas since to have drawled on and on would have perhaps proven somewhat boring. Space has been used wisely to discuss the details we felt were important.

Finally, we must thank the companies who provided us with the raw information and the individuals within them who took the time and effort to answer our questions.

ACKNOWLEDGEMENTS

Many friends and colleagues have helped to contribute to our understanding of IT implementation over the last few years and it is now time to officially thank these individuals. We pass particular thanks to the individuals and organisations that have provided case study material used in this book. In no particular order these people and organisations are: Graeme Chalmers at G2 Business Associates in Glasgow, Dr Sarah Bowden of Arup and Paul Gooding of BSRIA (part of the www.comit.org.uk project), Vanessa Ambler of Inspector Home Ltd, Ivelina Ivanova of pH Europe Ltd, Hilary Brown of the Chartered Institute of Building, Alan Hore of CITA Ireland, and two individuals and organisations who for reasons of commercial confidentiality wish to remain anonymous. They know who they are and we thank them for their contribution.

We also wish to thank Fiona Turner and Olivia Gill for putting together and formatting many of the diagrams and figures within the book.

Last but not least we must thank our families for providing love and support and putting up with the long nights and countless moods.

ABBREVIATIONS

AMS	Asset Management System
BPF	British Property Federation
CAD	Computer Aided Design
CAFM	Computer Aided Facilities Management
CBA	Cost Benefit Analysis
CCP	Change Control Proposal
CIOB	Chartered Institute of Building
CITB	Construction Industry Training Board
CRT	Cathode Ray Tube
DMS	Document Management System
ECDL	European Computer Driving Licence
EDI	Electronic Data Interchange
EDMS	Electronic Document Management Systems
EDRMS	Electronic Document Records Management Systems
ERMS	Electronic Records Management Systems
GUI	Graphical User Interface
HCI	Human Computer Interface
HWR	Handwriting Recognition
IAF	Impact Analysis Form
ICDL	International Computer Driving Licence
ICT	Information Communication Technology
IIMS	Integrated Information Management Systems
IMS	Information Management Systems
IT	Information Technology
JIT	Just in Time
KM	Knowledge Management
KPI's	Key Performance Indicators
KTP	Knowledge Transfer Partnership
LCD	Liquid Crystal Display
McE	Micro Enterprises
MOD	Ministry of Defence
NHS	National Health Service
OCR	Optical Character Recognition
OLED	Organic Light Emitting Diode
PDA	Personal Digital Assistant
PFI	Private Finance Initiative
PICA	Plan Implement Control Adjust
PPP	Public Private Partnerships
RAM	Random Access Memory
RFI	Request for Information
RFID	Radio Frequency Identification Devices
RIBA	Royal Institute of British Architects
ROI	Return on Investment

SME	Small Medium Enterprises
STEP	Social Political Economic Technical
SWOT	Strengths Weaknesses Opportunities Threats
USB	Universal Serial Bus
VOIP	Voice Over Internet Protocol

CHAPTER 1

The role of Information Technology (IT) within the construction industry

This chapter will consider the following:

- IT developments.
- The role of IT in construction processes.
- Project delivery.
- Plans of work.
- E-commerce and IT.
- A summary.

1.1 IT DEVELOPMENTS: SETTING THE SCENE

To deliver the broad range of construction projects that meet the demands of a wide range of clients at today's speeds and exacting standards means that the players involved in a construction project must fully utilise all available resources – Information Technology (IT) is one such resource. In today's business environment IT surrounds us and affects all that we do during normal day-to-day business operations and activities.

The sea change in the uptake and use of IT within the construction industry is evident from simple comparison of a project delivered 10 years ago and how a similar project is delivered today. Those involved in delivering Terminal 5 at Heathrow have embedded IT within the composite project delivery structures. The IT technologies themselves and applications software have helped drive this change: you can walk down any high street and buy low-cost computing power that year-on-year becomes ever more impressive. The range of software is extensive and what is not readily available off the shelf can be developed in a very short time, and usually at relatively modest cost.

A Personal Digital Assistant (PDA) or handheld computer (with prices ranging from under £100) now has the computing power that many a mainframe had only a decade or less ago. The majority of readily available laptops have such broad functionality that they become full scale work centres in their own right. When this basic computing power is coupled with Internet access, wireless operability and media enhancements, then it renders these machines as formidable tools when placed in the right hands. The proliferation of Internet use means that work can be shared any where on the planet – static locations are a thing of the past.

Whilst it is not the intention of this book to delve into the intricacies of what makes a computer 'tick' (it is assumed that the reader is familiar, or can familiarise themselves with such details), suffice to say that the processor and storage on modern machines has radically changed from those available ten years ago or even a few years ago. USB (universal serial bus) memory sticks (pen-drives as they are often called), CD's and DVD's now store in the order of Gigabytes of data and the chip speeds are also measured in Gigabytes: values only dreamed of a decade ago.

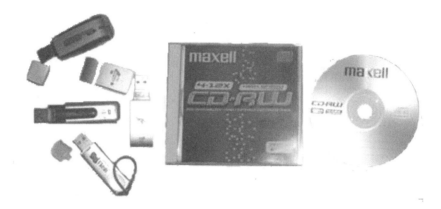

Figure 1.1 Images of typical USB sticks and CD/DVD

Where it is appropriate in the book to ally technical issues to the applications or solutions, then it will be in a manner which assumes that some background immersion in the technical aspects has already been undertaken. The applications software becomes ever more user-friendly and intuitive, so dispelling many of the inherent drawbacks that existed less than a generation ago. Graphical User Interfaces (GUI's) now provide the user with software interactivity which renders most applications almost foolproof. The preponderance of the ubiquitous Windows™ operating system has ensured that the proliferation of friendly systems continues unabated. Only 2 or 3 years ago few would even have hazard a guess that Voice over the Internet (VOIP) would become 'de-rigour'.

With this constancy of change in the underlying technology base and the ease of use of a plethora of applications, so the use of IT has expanded to encompass areas where previously they may been thought unsuitable due to complexity or operational reasons i.e. cost. This change in emphasis encourages us to establish a clear understanding of the nature and role of information technology and the part it plays in ensuring success at the both the project and organisation levels.

1.2 THE ROLE OF IT IN CONSTRUCTION PROCESSES

The functional role of IT can best be considered from three standpoints: the part it plays in each of the various stages of a construction project i.e. pre-tender, post-tender and post-completion. Also, there is a need to consider how IT affects procurement and the eventual routes adopted for each project and organisation. Finally, there is a need to consider the impact IT has on the humans involved in the delivery of the construction project.

Discussion on the implementation of IT in a construction project paradigm is necessary since the industry, by necessity, adopts approaches to problems that are different from those used in other industries: indeed the challenges faced by construction organisations differ from those encountered in other industries. Other industries predominantly deliver their product from a static production base, with a constant workforce and a supply chain which is uniform over significant periods of time. Construction does not benefit from these three key characteristics: its uniqueness stems in part from the nature of the industry, the flux in client types, the dynamism of the projects and the inherent nature of the organisations and the composite supply chain.

It may well be argued that construction differs from other industries in that governments (of whatever persuasion) have used, and continue to use, the construction industry as a regulator of the economy. Chapter 2 in this book will touch briefly on the 'multiplier' and 'accelerator' devices which bring about effects that have significant impact on all who operate in our industry.

In particular, the construction environment drives the need for innovation in approach to the final product i.e.

- Projects are basically one offs – each design is unique and seldom, if ever, repeated. Consider the situation where we are building 200 new homes. These houses on a large estate may appear similar, but they are not identical e.g. changing soil conditions may mean additional under-building, the buyer may wish different doors on the kitchen units etc.
- The clients are generally unique and their approach tends to be fairly individualistic. Even organisations who work for the same client on a number of occasions will find that no two projects, nor approaches to the projects, are identical.
- Profit margin reported within financial returns for organisations active within the industry show that there is a tendency for these reported profit margins to be low and there are considerable periods of lengthy negotiation prior to actually commencing operations on site. When the operations are under way the payment terms within the contracts often mean that cash-flow is negative for significant periods.
- Rapid start-ups are often demanded by clients in order to meet both their short-term and broader strategic needs. The contracting organisation has to respond even though the clients have delayed the commencement to suit their needs.
- The manufacturing content of most projects is fairly low. There have been a number of attempts at 'industrialisation' of the process: indeed Manubuild is one recent initiative which holds much potential to bring benefit to the industry (http://www.manubuild.org/). On larger projects there is a high degree of component delivery from manufacturing units, but these components are still assembled on site.
- With the diverse nature of the client base and the wide range of their changing needs, so the potential for repeat business is not always available. Therefore we are unable to reap the maximum benefits from direct learning experiences or process optimisation. This 'singularity' of project delivery differentiates us from a range of other industries which

may be similar, but are not the same e.g. shipbuilding, where the order is often repeated (even with a time lag between delivery dates).

- The industry structure is in some ways a real barrier to wider IT adoption. There are many players – some very small: the micro-enterprises i.e. businesses with less than 9 individuals within them, account for some 97% of all construction industry enterprises (NSO, 2005). These micro-enterprises often operate on very slim margins and the time commitment required to fully implement the most effective IT solutions is often at a premium. The barriers for entry to, and exit from, the industry remain low and so businesses come and go with alarming regularity.
- Workload for the industry and each constituent organisation is contingent upon a number of factors: each of these possesses the ability to act in isolation, and in conjunction with one another as components of a broader macro-system. With this multiplicity of possible work drive interactions, so the organisation seeking work has to be innovative and employ all resources to their best advantage.
- Those actually carrying out the activities on site are a predominantly transitory workforce, and practical skill focused. The industry continues to have a recruitment problem and those human resources required for the higher level positions are in short supply. The demographic 'timebomb' facing the industry can, to some extent, be mitigated by smarter use of IT.

1.3 PROJECT DELIVERY

Having said all of that, projects are still delivered on or below budget, safely, on or below the time thresholds set, and to the client's requirements. IT's part in this delivery can be seen at each of a number of stages within the overall 'Plan of Work'. There are a number of Plans of Work available for use, some more readily recognised than others e.g. Royal Institute of British Architects (RIBA), British Property Federation (BPF). It is not the intention to cover these in detail, but a brief synopsis of at least one Plan of Work provides a solid foundation for discussion on IT applications.

The RIBA Plan of Work is a robust process protocol which describes the activities from appraising the client's requirements through to post construction. The stages are also used in the appointment of the players within the project and contract documents used to secure timeous delivery of the works, and at the same time, to help clearly identify the services required of the various professionals and contractors.

The Plan of Work is widely recognised throughout the UK construction industry as a solid approach to modelling the framework under which any project can be delivered (Figure 1.2) it assists in the effective and efficient management of projects and forms the basis for numerous office procedures. The four main themes within Figure 1.2 illustrate how the elements within the Plan of Work integrate to provide a composite approach to project management and project delivery.

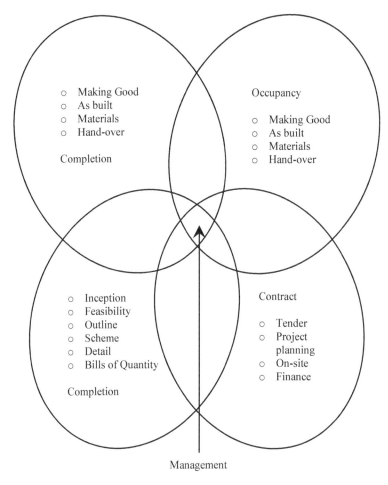

Figure 1.2 The typical project framework

1.4 RIBA PLAN OF WORK ELEMENTS (after the 1999 version)

A: Appraisal - The appraisal and identification client requirements and consideration of a range of possible constraints on development. This stage also includes the preparation of various studies aimed at informing the client whether to proceed or not and to engage in selection of the most likely procurement method.

B: Brief - Preparation of the strategic brief by (more normally on behalf of) the client confirming key requirements and constraints. At this juncture it is normal to identify the procedures, organisational structure/s and the various consultants etc. who will be engaged for delivery of the project.

C: Outline proposals - Development of the strategic brief into a full-scale project brief which now includes the preparation of outline proposals and some estimate as to the likely overall project cost. The chosen procurement route will be compared to others available and validated.

D: Detailed proposals - Complete development of the project brief and preparation of detailed proposals.

E: Final proposals - Preparation of final proposals for the project such that they will facilitate co-ordination of all components and elements that are to be incorporated within the project.

F: Production information - F1: Preparation of production information in sufficient detail to enable a tender or tenders to be obtained. Also, application will be made for the necessary statutory approvals e.g. Planning, Building Warrant.

F2: Preparation of further production information required under the building contract: generally comprising the balance of information required to complete the construction.

G: Tender documentation - The preparation and collation of various tender documents which should enable an interested contractor to tender for the construction of the project e.g. the drawings, the Bills of Quantities, the specifications, the contract document, the submission pro-forma, and other relevant materials.

H: Tender action - Identifying and evaluating potential contractors and/or specialists who are likely to be engaged in delivery of the construction of the project. Receiving and appraising tenders and submission of recommendations to the client for, informed, final selection.

J: Mobilisation - Letting the actual building contract: appointing the contractor, issuing production information to the appointed contractor and, arranging site handover to this contractor.

K: Construction to Practical Completion - Administration of the building contract up to and including practical completion of the project. This stage also includes execution and control of all site work & associated activities and provision of further contract documentation/information as and when reasonably required. Undertaking the 'snagging' and ensuring development of operating instructions, maintenance manuals, opening ceremonies, occupation, evaluation of facilities management and any staff training that is deemed necessary.

L: After Practical completion - Administration of the building contract after practical completion has been achieved, including: making final inspections and settling the final account/s. The project may be complete but now the occupier is faced with day-to-day living inside their edifice. IT plays a real part in the Facilities

Management of the building and even comes into play when considering demolition.

As stated earlier, it is not the intention to delve into the Plan of Work in significant detail at this point. For instance, there is no mention of individuals involved at each element of the plan of work nor detail on the tasks to be done. Hopefully these details will emerge as self-explanatory in the Case Reviews. Many of the elements within the Plan of Work lend themselves to delivery and management at a distance – the growth and robustness of the web allows such elements to be delivered as 'part and parcel' of e-commerce.

1.5 E-COMMERCE AND IT?

The question arises as to what does IT do for the Small to Medium Enterprises (SMEs), the Medium to Large, and the Large Enterprises? Buzzwords engulf us, and IT applied to the construction industry is unable to escape this phenomenon. E-business, e-commerce and e-technologies are buzzwords that we habitually read about in the press: removing the overlying hype gives us the chance to see what these terms actually mean to the enterprise? Even more importantly, what tangible benefits can they bring to the enterprise?

Internet-based technologies are tools that use Internet protocols to provide capabilities, which bring about gradual improvements in the way in which business is conducted within and out-with the enterprise. Table 1.1 indicates a number of the key features of each net approach.

It may be argued that the relatively slow uptake of these particular tools by the construction industry can be attributed to a general lack of appreciation of what is available and what can be achieved through the new and emerging technologies. As discussed earlier, the size of many of the enterprises can be seen as a barrier and yet their smallness should provide an ideal breeding ground for innovative, productive, approaches to problem solving.

Managing a project over the Internet, building a computer-mediated tendering system, purchasing and selling materials online, automating transactions and the composite supply chain, are not examples of dreams: they are realities now and demonstrate ways in which innovative, forward-looking, companies in the industry are exploiting the power of the Internet.

Table 1.1 Properties of net approaches

	INTERNET	INTRANET	EXTRANET
Users	All	Employees	Selected groupings
Information	Fragmented	Enterprise specific	Secure sharing
Access	Public	Private	Semi-private

Many within the industry may see the greatest barriers to introducing e-driven change in the construction industry as being:

- Slow-decision making – only when it suits those making the decision, at other times the decision making process can be extremely swift. The fear of something new and unproven hangs in the air waiting for others to prove the method.

- Bureaucratic organisational style – many overseas buyers are active in the market place and this will stimulate much needed change.
- Shortage of competent IT staff – those IT staff with some immersion in the industry are rare indeed, but there is no excuse for not taking I the uninitiated and bringing them 'up to speed' in-house.
- Lack of financial resources – in an industry that continues to have strong demand a more pragmatic, long-term approach to finance is required.
- Difficulties in predicting Internet behavioural patterns of the market – one of these 'nuts' that effective marketing can, and does, overcome.

Yet starting out on the e-business path is now straightforward and relatively low-risk: a company can host a web-server internally or outsource it and either customise its functionality or have it customised to fulfil their corporate needs. Readily available web-enabled and web-based software can be customised into enterprise specific systems.

IT and e-systems can bring improvement to the UK construction industry through eliminating well-documented inefficiencies within the overall construction processes e.g. duplication of information, poor filing and data retrieval, multiple sources of similar data and cumbersome material purchase processes, project management, procurement, tendering, modelling, and knowledge management.

Purchasing of services/contracts/materials is still a predominantly manual process based on e-mail, phone, and fax. The resulting high inefficiencies and inadequate controls ensure there are significant opportunities for cost savings. E-procurement solutions can streamline all processes and offer significantly reduced costs to the enterprise.

Project management has improved through using tools known as 'project collaboration' (mainly project extranets). These create a construction project data environment on secure servers and are accessible to those selected members of the project team. They can view, upload, amend and discuss many/all documents, correspondence and drawings, dependent on access rights assigned to the individual. The resulting transparency and availability of the data in real-time makes these tools extremely powerful – the resultant audit trail is clear and mitigates any reliance on a blame-culture.

1.6 SUMMARY

In IT terms, it may be argued that the only constant is change. This IT change can, and should, be harnessed as a useful resource and applied to the numerous challenges and problems that are faced on any construction project. The hardware configuration and software applications most suitable are issues to be decided by the enterprise since they understand the direct issues to be resolved.

What management of the enterprise must do is consider the range of options open to them and appraise the benefits to be gained from implementing a chosen route. Cost-benefit analysis is not rocket science, rather it demands that the management sit down and consider in a holistic fashion what they are seeking to resolve, if the route they have chosen is the best for their enterprise, how it will be implemented, what the effects on the enterprise will be and, the benefits that will result from the use of IT.

At the end of each chapter we have included some questions for you to contemplate as pointers (questions to ponder). They expand and relate to, the future direction/influences found in each of the chapters.

The rest of the book

Chapter 2 focuses upon the industry drivers that underpin IT and its implementation. These drivers are considered from a number of perspectives in order to deliver the 'holistic' view. How IT influences productivity and efficiency of resources is investigated along with the use of human resources in IT projects.

Chapter 3 considers in some detail the interaction between humans, computers and IT applications in general. Attention is given as to how the working environment affects and is influenced by IT.

Chapter 4 reviews the need for IT within construction organisations and examines issues such as the business processes, information management, conflict, and decision making with specific emphasis being placed on how IT can aid the day-to-day operations of construction organisations.

Chapter 5 examines the capture of knowledge within the construction industry in general with particular attention being placed upon the effect of knowledge within the wider supply chain.

Chapter 6 considers the use of information within the construction industry and also at the project level. The reliance upon traditional paper based methods of communication is discussed at length together with current site based methods of data capture.

Chapter 7 looks at IT Tools available for adoption within the construction industry including Electronic Document Management Systems (EDMS), Extranets and Mobile communication devices. Discussion is given to their role and how they are implemented within the construction industry.

Chapter 8 examines the implementation process. Discussion is provided into current working processes and also the quality systems behind implementation such as ISO 15489 and ISO 17799.

Chapter 9 looks at IT implementation on 3 distinct construction projects and 8 organisations with specific problem areas. It considers the problems being faced and the implementation of the IT solution. Our thanks go to each and every organisation that provided the case reviews for us. Without their support the book would perhaps have ended up as just another tome droling on about the theories.

Chapter 10 provides an overview of the preceding chapters and considers what has been learnt from the implementation of IT on the 3 distinct projects and through case reviews of the systems being utilised for specific purposes. The Do's and Don'ts are provided as a means for others to grasp where they might be heading and ensure that their experience is enjoyable.

CHAPTER 2

IT drivers within the industry

This chapter will consider:

- The operation of the industry and its structure from a project perspective.
- General economic data e.g. size, GDP, types of work, range of clients, etc.
- The 'multiplier' and 'accelerator' as it affects the industry.
- The client side, demands, expectations, briefing.
- Business processes, systems integration, and training of staff.
- Client IT sophistication (existing IT systems).
- The suppliers (all suppliers including IT) needs, demands, expectations, briefing, business impact, systems integration, training of staff, forward looking, cost implications.
- The subcontractors' point of view demands, expectations, briefing, business impact, systems integration, training of staff, forward looking, cost implications.
- The need for joined-up actions i.e. taking the holistic view, the use of holonic networks and the impact of business syntegrity.
- A summary.
- Questions to Ponder.

These diverse drivers are considered from a number of perspectives in order to deliver the 'holistic' view. How IT influences productivity and efficiency of resources utilised within the industry is investigated along with the use of human resources in IT projects.

2.1 OPERATION OF THE INDUSTRY FROM A PROJECT PERSPECTIVE

The construction industry, as has been argued in the earlier chapter, is unique: unique in the sense that we deliver projects through teams which are formed for that specific project's duration and then when near completion, we dismantle this successful team. The structure of the industry is such that we rely on micro-enterprises (an organisation with less than 10) to actually deliver the project outcomes. Indeed the structural figures available show that some 98% of our industry is comprised of micro-enterprises. With this in mind the next few pages will focus on the micro-enterprises and then pan out to the wider industry level.

2.2 WHO OR WHAT ARE THE MICRO-ENTERPRISES?

We all may well be aware of the terms Small–Medium Enterprises, yet many are not fully able to identify their underlying characteristics. The Bolton Committee (1971) in its Report on Small Firms gave what may be seen as the principle definition of a small firm: a small firm is an independent business, managed by its owner or part-owners and having a small market share.

The Bolton Report provided a number of statistical definitions which aid in distinguishing the SME's from other enterprises. The report also considered size in each industry as being relevant to that industry i.e. an enterprise could be small in relation to one industry where the market is large and there are numerous competitors; whereas, an enterprise of similar characteristics could be considered large in another industry which has fewer players and/or generally smaller firms operating within it.

Across UK Government, size is measured according to numbers of full-time employees or their equivalent. For statistical purposes, the UK's Department of Trade and Industry usually employ the following definitions:

- micro firm: 0-9 employees
- small firm: 0-49 employees (includes micro)
- medium firm: 50-249 employees
- large firm: over 250 employees

At the broader European level, the European Commission have established a single definition of SME's which is outlined in Table 2.1. The term Small to Medium Enterprises (SME's) embraces 99.78% of all enterprises found within the UK economy (NSO, 2002). The SME cohort contains a sub-group known as micro-enterprises: enterprises comprising less than 10 individuals which account for 94.37% of all enterprises in the UK economy.

Table 2.1 European approach to SME definition (Adapted from
http://www.dti.gov.uk/SME4/define.htm)

Criterion	Micro	Small	Medium
Maximum number of employees	9	49	249
Maximum annual turnover (Euros)	-	7m	40m
Maximum annual balance sheet total (Euros)	-	5m	27m
Maximum % owned by one or jointly by several enterprise(s) not satisfying the same criteria	-	25%	25%

Within the construction sector, micro-construction enterprises (McE's) account for almost 98% of the total number of all private contractors; engage some 44% of all employees, and account for almost 33% of the value of all work done (NSO, 2002). These statistics suggest that there is a need to fully comprehend the management of these enterprises and how they engage in all facets of project management and delivery. Management practices within the broad grouping of small to medium enterprises (SME's) has received much attention over the last decade and yet little has focused upon the real engine house of the construction industry, micro-construction enterprise (McE).

Good management skills have been identified as a key factor in the performance of the economy as a whole and the effective operation of individual firms. It has been recognised across government that there is a need to improve the practice of management in the UK and the training and development processes underpinning this aim. However, a fundamental research gap exists which is associated with management development within the context of micro enterprises and in particular with McE's.

Table 2.2 shows the general industrial picture for the number of enterprises, employment and turnover in the private sector by size of enterprise. The significance of these contributions provides a powerful argument for the drive to have them included as a separate and specific entity within policy formulation and implementation.

Table 2.2 Number of enterprises, employment and turnover in the private sector by size of enterprise (Adapted from Annual abstract of statistics, National Statistics Office, 2002)

All Industries	Number of Enterprises		% of all Employment (000s)		% of all Turnover (£million)
All enterprises	3,746,340		22,622		2,112,013
With no employees	2,596,395		2,888		152,383
Employers with					
1-4	747,655	Micro =	2,230	Micro =	172,637
5-9	200,320	3544370	1,436	6554	122,750
10-19	112,695	=94.6%	1,570	=29%	148,725
20-49	54,845		1,686		168,840
50-99	18,130		1,255		141,269
100-199	7,905		1,104		135,171
200-249	1,620		362		42,791
250-499	3,245		1,128		157,237
500 or more	3,530		8,964		870,211

However, a number of general economic reviews and specific construction industry reviews e.g. Latham, 1994 and Egan, 1998, undertaken with the premise of informing the development of management education and learning have largely ignored the micro-enterprise context. This omission becomes even more complex when considering IT in the micro-enterprise. We arguably rely on them to deliver project success and yet in many instances they are at the bottom of the IT implementation 'food chain'.

If the figures on construction specific micro-enterprises are stripped out from the data then we see a clearer picture in terms of their mass, scale of operation and impact on the overall industry. Table 2.3 shows specific data relating to construction industry micro-enterprises.

Table 2.3 Construction industry data (Adapted from Annual abstract of statistics, National Statistics Office, 2002)

	No. of Enterprises	As a % of all enterprises	Employed (000's)	As a % of all employed	Turnover (£ Million)
All enterprises	691,800		1,666		136,927
With no employees	571,455				25,546
Employers with	Micro Total = 675315, =97.6%		Micro Total = 976000 =58.6%		
1-4	85,445		244		19,406
5-9	18,415		130		10,659
10-19	9,895		135		11,834
20-49	4,535		136		12,900
50-99	1,200		81		9,333
100-199	470		64		10,628
200-249	80		18		1,958
250-499	165		58		7,202
500 or more	140		200		27,460

Those charged with the responsibility of developing policies to support management and wider workforce development activities have at best assumed that the micro-enterprise context is not unique and therefore can be embraced within the term Small and Medium sized enterprise. However, SMEs are generally seen as enterprises employing up to 249 and in many ways are a world apart from the true micros. The less formalised and more personal management practices found in micro-enterprises often ensures that they remain distinct from their relatively larger counterparts (EC, 96).

2.3 MICRO-ENTERPRISES IN THE UK CONSTRUCTION PARADIGM

Any discussion of micro-enterprises within the construction context needs to draw attention to their heterogeneity. When discussing micro-enterprises the term should be seen as an inclusive paradigm which embraces start-up enterprises, self-employed managers with one or two employees, owner-managed businesses, team-managed businesses, family businesses, ethnic businesses, and businesses with a plethora of differing legal status. Given the ease of entry to the construction industry, the cohort title must by necessity, be suitably ambiguous. The concepts are not necessarily mutually exclusive and can be used interchangeably, with the caveat that this can, and indeed does, lead to a lack of clarity in policy discussion.

Micro-construction enterprises (McE's) operate as complex social organisations which engage in a range of holonic networks across a broad geographic arena. The managers within the McE's have a pivotal role in shaping the intrinsic culture and any transition towards change. In mature enterprises, management practice will have developed over a number of years and external drivers which seek to change these established practices, may be viewed with a degree of uncertainty (Griffith & Headley, 1998).

2.3.1 Management skills in the micro-construction enterprise

Those managing micro-construction enterprises require the possession of a diverse range of skills which centre on a set of core competences comprising: functional or task based skills (such as marketing, accounts and administration); strategic, analytical thinking and planning; and, people skills. These skills being put into action both within and out-with, the enterprise's boundaries. To operate effectively as a manager of the micro-construction enterprise, the individual will require a broad range of abilities encompassing narrowly defined skills through to understanding personal behaviours and attitudes (Lingard & Holmes, 2001).

These skills can be represented as shown in Figure 2.1. The Construction Industry Training Board (CITB, 2002) argues that irrespective of the size of construction firm, the management team must excel in: winning contracts; managing the delivery of those contracts; planning the future of the business; and, managing employees, customers and suppliers. The justification for the CITB argument revolves around four processes that construction companies need to be good at i.e. *the key processes*. The construction enterprise must:

- Have a strategy that is based on reality and exploits the market – *strategic management.*
- Successfully tender for, and win, profitable business – *business development.*
- Manage the construction process to ensure the job is on time, to specification and
- Makes a profit – *construction management*
- Maximise the value received by their stakeholders (clients, employees and suppliers) – *stakeholder management.*

The fundamental 'Know-how' must be supplemented by 'Know-what' (knowledge), 'Know-why' (understanding) and 'Know-who', the ability to cultivate and learn from the experiences gained within the holonic networks and the external relationships. These abilities and skills needs will change as the enterprise develops or its business paradigm metamorphoses and as the constraints acting upon the micro-enterprise realign.

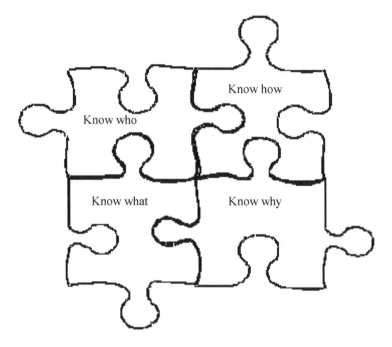

Figure 2.1 The micro-enterprise skills map

The general view is that micro-enterprises have to focus on the short-term and immediate survival, so undermining an appreciation of longer term benefits to be derived from training. Past research and a variety of reports have shown that training is at best reactive, targeted at specific tasks/problems facing the business, and fire fighting in nature i.e. specific solutions to specific crises (CITB, 2002).

The old adage that 'time is money' applies directly to the micro-enterprise and many micro-enterprises find it difficult to spare time for managers or key workers to engage in training (even though a broad range of professional bodies require specific amounts of continuing professional development in order to maintain individual membership).

Further constraints include: cost, a major constraint; location and timing of events i.e. unsocial or peak business hours; the range of training available; the accessibility of the training i.e. often pitched at the wrong level and, the perceptions of training needs, the capabilities and attitudes of owner-managers, and the quality and relevance of training offered; perhaps the key constraint is the level of importance attaching to management skills themselves (Thomson & Gray, 1999). The CITB (2002) see the largest areas of growth within the industry occurring in the Managerial, Clerical, Technician and Professionals occupation categories. Selections of the actual employment levels for 1999 (and 2005 projections) are shown in Table 2.4.

Table 2.4 Employment structure, Adapted from: CITB's Employment and Training Forecast 2002 (SB Specialist Building; CE, Civil Engineering)

Occupation	Number employed in 1999	Projected employment in 2005	Average annual growth (%)
Managers	133600	145400	1.1
Clerical	117800	128800	1.4
Professionals	38200	41500	1.8
Technicians	37800	41700	1.3
Carpenters/joiners	200900	211500	0.4
Bricklayers	113200	117800	0.0
Painters	84500	85700	0.4
Steel Erectors/Structural	13800	14300	0.2
General operatives	73200	75700	0.6
Electricians	130800	139200	0.5
Plumbers	106300	113800	0.6
Total	1386400	1459600	0.5

Therefore, the actual skills needed by these incoming individuals, to manage the micro-construction enterprise range from those seen as primarily financial to those based on interaction with humans. The Small Businesses Skills Assessment (SFEDI, 2002) draws upon a number of sources when establishing their basket of skills needs and how these skills impact upon the enterprise. Table 2.5 summarises the skills requirements. Homan *et al.* (2001) argue that customer care, cash flow/ credit management and sales are the principal management techniques thought most likely to contribute to the success of micro and small enterprises: areas where appropriate IT can ably assist a business.

Within Table 2.5 it is possible to identify management skills that centre on internal operations of the micro-enterprise and skills that relate to the external business environment. The internal skills basket requirements will be influenced by the nature of the enterprise and by its very size. As noted in Table 2.3, 152,383 UK construction enterprises have no employees. Clearly, employee related management skills would be of little relevance in this context. It reinforces the argument that small businesses cannot be treated as homogenous in terms of their skill needs. Whilst the lists of skills and abilities can be applicable to many small businesses they should not be taken to be of universal relevance or of equal importance in all contexts.

Table 2.5 The skills needs distribution

Skill area	Specific tasks	Impacts
Marketing	Planning sales improvements, Exploring and capturing markets, Converting sales leads	Overall business activity levels.
Customer relations	Looking after customers, Improving service quality to customers, Aligning service with customers	Image and operation of the enterprise.
Directing and Controlling	Business planning, Service portfolio management, Implementing systems	The delivery of the service and its link to customer's needs.
Location	Premises, Utilisation of equipment, Access to markets	The ability to access customers and deliver services efficiently.
C&IT	Developing interfaces/equipment, Communicating with IT, Trading with IT	The wider operation of the enterprise and its responsiveness to customers.
Financial management	Improving the enterprise's finances, Securing lines of credit, Reducing debtor balances	Stability of the enterprise and perception among peers.
Time management	Delegating tasks where appropriate, Installing appropriate structures, Balancing work/life needs	Freedom to engage in the broader strategic issues facing the enterprise.
Human resources	Reviewing staffing balance, Ensuring performance, Maintaining the right staff	How the service is delivered.
Resource management	Ensuring full resource utilisation, Securing service levels from others, Modelling the resource base	Getting the most from the internal strengths of the enterprise.
Health and Safety	Risk management, Executing assessments and audits, Managing and reducing dangers	Ensuring that services are delivered safely and that humans are cared for.
Strategic management	Reviewing direction, Developing new plans, Looking forward and taking decisions	The long-run survival of the enterprise.

The skills set and outlook of managers within micro-construction enterprises are critical in setting the vision and culture of the organisation and determining the extent to which other staff have the opportunity for training and development. The industry may well voice concern that attitudes to training in SME's are primarily conditioned by the attitudes and experiences of the owner-managers (Wyer et al., 2000).

The micro-enterprises do engage in delivery of the industry's overall turnover and as such contribute to the broader economic base of the industry. The construction industry accounted for over £57Bn in 2004, encompassing works as diverse as small domestic extensions and repairs to Terminal 5 at Heathrow Airport. The works are generally divided into Public and Private sector although this historic division is being eroded by issues such as Public Private Partnerships (PPP). Within this funding division there are further sub-divisions of new-build, refurbishment and maintenance. Each of these particular areas has specific demands on IT and at the same time utilise generic products e.g. word processing.

2.4 THE 'MULTIPLIER' AND 'ACCELERATOR'

Pick up any economics book of any substance on our industry and you will come across the terms 'multiplier' and accelerator'. Each of these terms has considerable importance to those contemplating the implementation of significant IT.

The 'multiplier' is concerned with the action of investment in a construction project: think of it as a demand generator. If a client comes along with £1Mn to invest in the development of a block of flats this demand in turn fuels other demands. The flats require to be designed, they require specialist services e.g. heating & ventilation, they require professional cost and project management services, they of course require the contractor to bring plant, labour and materials to the site in order to carry out the construction works. This generation in business sparks demands on suppliers and other sub-contractors so increasing the effect of the original investment. Governments have long seen this effect as being useful in stimulating the economy when it may appear sluggish. Pubic works bring about a significant increase in employment and inject vital sums of cash into the broader economy.

The 'accelerator' can best be viewed as the throttle on a car. To increase the movement within the economy and so business, governments adjust interest rates up or down or inject cash, to increase or slow the demand side on the economy. With historically low interest rates for some considerable period of time we have witnessed unprecedented demands for construction activity – be it new homes, commercial or industrial buildings. The statistics on industry output shown in Table 2.6 and Figure 2.2 illustrate the growth in output values for the industry over the last decade or so.

Of course a look at the broader issues surrounding our industry would not be complete without some consideration of the client side. The client side is generally thought of as being the private sector in all its guises and yet for the purposes of reviewing client impact on IT we have to take a much broader view and see the client side as all those who commission construction activity. The reason for this can best be seen if we consider a Local Authority who wishes to implement an

Asset Management System (AMS). The activities generated within the AMS have direct impact on all built assets owned by the Local Authority and will in many instances see the maintenance and repair activities carried out by an in-house department e.g. Direct Works. Whist the argument may be made that there is no external client, there is an internal client.

Table 2.6 New works order values (Adapted from Construction Market Intelligence, Department for Trade & Industry)

New work (£M) for Public & Private sectors

Year	New housing	Infrastructure	Other	Total
1994	7,407	5,075	12,304	24,786
1995	7,128	5,594	13,669	26,391
1996	7,004	6,311	14,432	27,747
1997	7,971	6,301	15,505	29,777
1998	8,423	6,170	17,743	32,336
1999	8,407	6,187	20,874	35,468
2000	9,977	6,441	21,122	37,540
2001	10,219	7,146	22,467	39,832
2002	12,077	8,077	25,070	45,234
2003	15,354	7,363	27,505	50,222
2004	19,425	6,490	31,177	57,092

Clients then have a range of demands and a set of expectations that must be communicated accurately to those delivering the construction services. Briefing has to focus upon attaining the best possible business impact and requires that IT systems used within the project or service boundaries are fully integrated. Integration of course relies on commonality of purpose, the provision of appropriate training for all staff engaged in operation of the systems and in looking forward towards completion of the tasks and hence future work activities. There are cost implications which must be borne in mind: the client's IT systems may be readily accessible, they may however require adjustments on both sides to accommodate and degree of interoperability.

Client IT is becoming extremely sophisticated as they engage more and more specialist service providers to deliver to them systems which accommodate their needs. Interoperability, as an integral factor within the design of any IT system, allows sophistication to be embedded and to some extent included future proofing within the system. The issue of 'legacy' (allowing outdated systems data to be retrieved and accommodated within new systems) also has to be considered from the client's point of view. They may have old systems operating which they need access to at some specific points in time.

The same caveats apply to the various suppliers we use in delivering construction activity. Although the suppliers tend to be generally more active in providing an outward looking access point for potential customers to access their servers and systems.

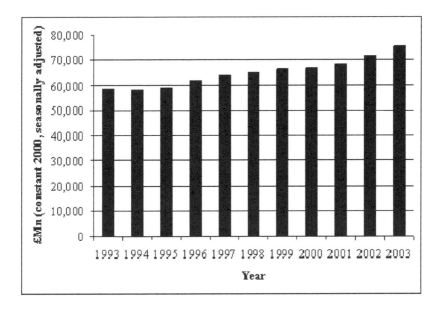

Figure 2.2 Output growth over an 11 year period

The construction industry delivers its successes through a range of complex supply chains. At the root of these chains are the sub-contractors found operating at the 'chalk face'. The supply chains are often long and regulated by conditions of contract which in many instances are not conducive to innovative working practices. And yet, the use of a range of IT is designed to engender a spirit of innovation in the project team. Project extranets and the like have broken many of the traditional barriers to access and sharing of information, therefore, the contractual arrangements also have to have inbuilt innovativeness in order that the team can maximise opportunities.

The sub-contractors (generally assumed to be SME's) have needs that drive their systems design and implementation which are often time critical events for them. The need to have the IT in place and fully operational affects the business processes and can have significant impact on the profitability of the business.

The sub-contractors tend to be keen participants in the implementation of novel ideas within the industry and look to the large organisations to engage them fully in developing IT systems that will operate across organisational boundaries.

There is clearly a need for joined-up action from all parties to the construction project. The holistic view is taken within this book, since only from that standpoint can we appreciate that there are a range of players and each has slightly differing IT requirements. Figure 2.3 illustrates the complexity of interactions on any project, emanating from the interaction of the three axes themselves (e.g. process with functions) and from the sub-divisions within each of the axes individually. As an example, if we consider how 'size of firm' influences IT design and adoption within an organisation then we begin to appreciate how complex the task of actually delivering project success is.

Taking other factors from each of the three axes and adding them together provides us with a range of possible scenarios that those charged with managing, and delivering, the project must cope with. Some of these are readily controlled, others require a much broader appreciation of underlying forces e.g. interpersonal relationships must be engineered and managed such that conflict is avoided whilst the project is under way. The issue of how entrepreneurs react to, and under, such 'controlled' environments adds further tensions to the project management. IT very often becomes a tool which is used both for benefits and against the interacting parties.

Added to this complexity is the planopy of holonic networks found within the industry. The holonic networks formed for delivery of the project ensure that complexity is an integral part of our delivery mechanisms. Holonic networks are usually short-lived networks of relationships formed for a specific purpose i.e. delivery of a construction project. They change with the project, the supply chain adopted, the project team members and also, with the demands placed on the composite team e.g. the delivery of a new school does not require the same team as would an airport terminal. Nonetheless what is required is that the team adopts a common approach to success of the project and so engages in shared (or at least communal) business processes and engenders a spirit of team syntegrity.

These drivers must be considered from a number of perspectives in order to deliver the 'holistic' view of project success. How IT influences productivity and efficiency of resources is key to understanding IT's role in bringing about success and also its role within the broader organisational framework. The human resources within the organisation employ IT to aid them in securing targets set: the exact scale and mode of employing IT of course varies with the organisation and the target.

2.5 PROCESSES, FUNCTIONS AND CONTINGENCIES

The range of processes which can be impacted by an IT system are shown on the left-hand vetrix of the triangle in Figure 2.3. The management of a number of the functions is attempted within some current IT systems, but very few systems attempt to manage or integrate all of the functions.

Legal processes are still held within the realms of specialist individuals, even though much of the background knowledge could be derived from and IT based expert system. Financial management via IT systems are routine, but still poorly done: very often the end-user is left out in the design of the financial management system and the desired output is missed. IT based administration has been well developed over the years and yet full integration and hence a composite system still seems to elude the industry. The remaining processes are centred on the 'soft side' of business and a number of human resource systems have been developed which offer significant benefits for the end-user.

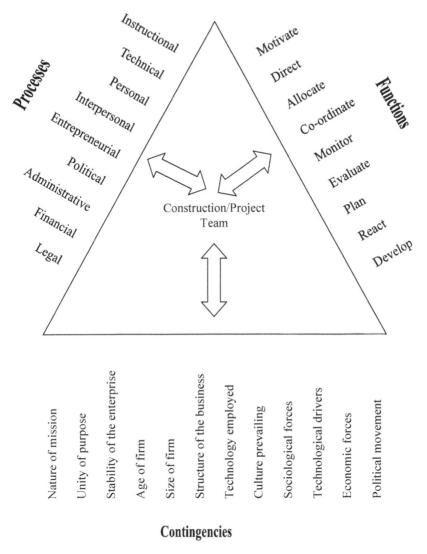

Figure 2.3 The drivers acting on the project team members

2.6 SUMMARY

The industry may be considered to be somewhat of a paradox: it may be argued on the one hand that it is amorphous and yet, clearly has a structure. The micro-enterprises who account for the vast majority of active companies found within the industry are self-reliant entities who only owe allegiance to the next contract holder. Their ability to integrate into a number of supply-chain configurations ensures that the holonic networks formed are transient and capable of rapid realignment. The micro-enterprises have IT requirements which in many ways mirror the larger enterprise and at the same time have to be tailored to meet the needs of a very small business. The supply chains found within the industry are only now beginning to implement shared IT across a number of platforms, to the benefit of all involved in the project.

The micro-enterprises operate a range of processes and undertake functions which do not directly map onto the larger organisations and yet the industry expects seamless integration of IT systems. The range of drivers impacting on how the business (of all sizes) operates is numerous and those managing the businesses must be aware of the potential force from each of these drivers. The industry has shown considerable growth and the projected levels of demand for construction services look buoyant for the next decade. The consideration of economic forces shows that the industry is important on the national scale and the part played in the regulation of the UK economy and the wider global market cannot be underestimated.

Having considered some of the broad range of forces acting upon the industry it is important to remember that the organisation – be it micro or large – relies on its people to carry out the range of duties required as integral parts of project delivery. These people form the core of the industry and it is they the focus falls upon in the next chapter. However, before moving on to the next chapter some questions and points for you to ponder.

2.7 QUESTIONS TO PONDER

- Given the current high levels of demand for services within the industry, will we see an increase in the number of micro-enterprises and therefore an increase in the number of players on any one project, with resulting issues over communication and information management?
- Profit may be held to be the key driver for the industry: if so what impact will changing IT systems have on the profit levels expected?
- The UK construction industry has a high demand for skilled operatives: will we see the day when semi-skilled individuals will, via the web, up-skill and deliver a service they were not initially trained for?
- Integration of IT systems relies on the end-user having a say in the initial brief and design of the IT system, why are there not more IT expert systems which facilitate extraction and development of the user's brief?

CHAPTER 3

People and IT systems

This chapter will consider:

- Human aspects of IT-Human–Computer Interfaces (HCI).
- Stress and other personal matters.
- Training for IT use.
- A summary.
- Questions to Ponder.

3.1 HUMANS AND IT IMPLEMENTATION

IT in any organisation relies on humans: the software and devices have not yet reached the stage in their development where they can fend entirely for themselves. For effective operation of IT, the humans have to engage in an appropriate manner. This interaction is contingent upon a number of factors as highlighted in Figure 3.1. The organisation must consider the impact of e.g. the hardware configuration, the operating system and software bundles, the skills of the humans and work environment. A number of these factors will be discussed in some detail to provide the reader with the fundamentals which underpin a solid view of the composite IT in action.

How humans interact with software and a range of devices has almost become a science in its own right. However, due to space and time constraints, there is no intention to discuss this 'science' in the depth that may be demanded or justified by others. A solid overview and resulting appreciation is all that is intended.

3.2 HUMANS AND COMPUTING DEVICES

Understanding HCI demands that we consider five interrelated aspects: the nature of human-computer interaction, the use and context of computers, human characteristics, computer system and interface architecture, and the development process.

Searching for a precise definition of the term Human-Computer Interface (HCI) quickly leads to the conclusion that there is currently no widely agreed definition of the term and in fact what emerged from the search was a view that the extent and range of topics which coalesce to form the area of HCI was broad and interdisciplinary. As a discipline in its own right it can be seen to be concerned with the design, evaluation and implementation of interactive computing systems for human use along side the study of major phenomena surrounding them.

HCI draws on several disciplines, each with different emphases e.g. computer scientist are concerned with the application design and engineering of human interfaces, psychologists are interested in the application of theories of cognitive processes and the empirical analysis of user behaviour, the sociologists are focused upon interactions between the technology, the workplace, and the organisation and, industrial designers look to comprehend the interactive products themselves.

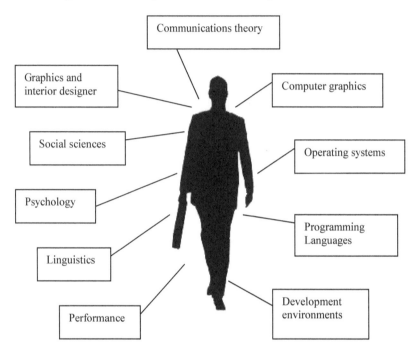

Figure 3.1 The wider influences on HCI

From a computer scientist's perspective, it rightly could be argued that the focus should be on interaction and more specifically on interaction between one or more humans and one or more computing devices/machines. Given the scale of IT's implementation within larger organizations, this focal point becomes an important aspect in the overall deployment of IT. Workstations are no longer isolated boxes: with increasing wireless connectivity they mesh seamlessly with the broader organization and its subsystems e.g. lighting, desk locations, working times, etc.

There are a number of other disciplines with points of view that would place the focus of HCI differently from that perceived by the computer scientist. HCI should be seen as an approach which melds an interdisciplinary area. For the sake of clarity in this book, HCI is taken to be 'that area of study concerned with the joint performance of tasks by humans and computing devices'. By necessity this means that we have to consider and embrace a number of cogent drivers as shown in Table 3.1.

This umbrella approach then allows for the inclusion of science, engineering, and design aspects within the broad title of Human-computer interaction. The design of many computer applications invariably requires the design of some individual component of the system to accommodate interaction with the end-user and their work environment. It is implicit in good systems design that the designer understands how to arrive at the functional requirements of a system, how to disseminate this functionality to the end-user, how to build the system, and how to test the designed system in the field.

Since HCI focuses upon human/s and computing device/s in communication, it wrings supporting knowledge from both the human and device sides. On the human side, communication theory, graphic and industrial design disciplines, linguistics, social sciences, cognitive psychology, and human performance are all relevant. Engineering and design methods are relevant and this importance is reflected on the device side by considering techniques in computer graphics, operating systems, programming languages, and development environments. This interaction map is shown in Figure 3.2. Comprehension of how these issues relate to each other helps in the development of a robust and meaningful appreciation and understanding of human-device symbiosis.

Table 3.1 Drivers of system detailing

Driver	Influence
Communications	The structure of communication between humans and devices.
Human activity	Human capabilities to use the device (+functionality of interfaces).
System processing	The underlying algorithms and programmes driving the devices and interfaces themselves.
Design	The engineering concerns that arise in designing and building appropriate interfaces.
Specification	The process of specification + design + implementation of interfaces.
Trade-offs	The resultant design trade-offs.

The means by which humans interact with computers continues to evolve rapidly and the reader should be aware that human-computer interaction is, in the first instance, affected by the future forces shaping the broader computing paradigm. Some of these forces are considered in Table 3.2. Which force will be the stronger/st remains to be seen: what is certain is that those implementing IT must take cognizance of the forces and accommodate them within their systems.

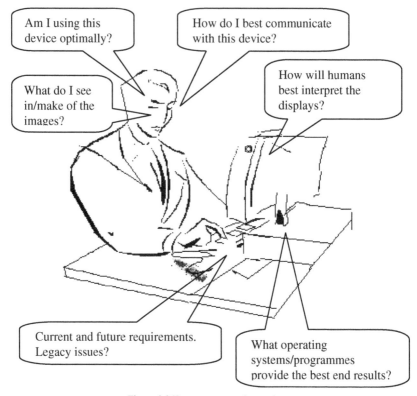

Figure 3.2 Human-computer interaction

Based on these forces and underlying trends, it can reasonably be expected that the future for HCI holds potential for development of IT systems which draw on 4 broad areas in order to provide the optimum solution for the end-user. These broad areas being: system transmission rates, visual impacts, pervasive computing, and end-user shaping. Implementation of some (or all) of a number of characteristics will bring about changes in the way we see and interact with IT.

Many of these future forces rely to some extent for their progress on increasing transfer rates. When considering system transfer rates we must be aware of:

- Ubiquitous communication. Computers are increasingly able to communicate through a range of high speed local area networks (office LAN systems currently have speeds in the order of 100Mbps), nationally over wide-area networks (with similar speeds), and via a plethora of wireless approaches which include infrared, ultrasonic, cellular, and other technologies e.g. Bluetooth. Data and other IT services are readily accessible (some free, most to be paid for) from many public places through 'Hotspots' e.g. at airports, bars, etc.

Table 3.2 Future force trends

Future force	Impact on computing
Decreasing costs of hardware leading to larger memories and faster systems.	The relative cost for high speed and capacity processing reduces. RAM values and hard-disks regularly store in the order of 80Gb. USB flash devices (sticks) are available with storage up to 4Gb.
Reduction in power requirements and miniaturisation of hardware leading to portability.	The decrease in size, and increase in functionality of portable devices has meant that the static office and workstation are relics of the past.
New display technologies that bring about novel packaging of devices.	Head-up displays enabling multi-tasking; decrease in power requirements of screens, brings about a radical reduction in the dimensions of devices.
Assimilation of computing devices into the wider environment (e.g., Digital televisions, DVD recorders, VCRs, numerous kitchen appliances e.g. (microwave ovens).	Many domestic appliances now have significant computing power and specific applications packages built-in at the factory (often coupled with other technologies e.g. RFID.
Increased development of network communication and distributed computing.	Wireless support for a broad range of devices (e.g. Bluetooth and Infrared) means that these devices are freed from a single location i.e. the desk.
Increasingly innovative input techniques (e.g. voice, pen), combined with lowered cost, leads to rapid uptake of IT by people who may have been excluded or placed on the periphery.	The Anoto pen is a specific example of the change in input approaches – the Anoto being one example of spatially aware input devices which enhance the functionality of the IT system.
Wider social concerns leading to improved computer access by currently disadvantaged groups (e.g. young children, the physically/visually disabled).	Disability legislation driving new interactivity modes.

- High-bandwidth interactivity. The rate at which humans and their devices interact has continued to increase: in part due to the changes in the speed of the computer chips themselves and also the enhancements in modern graphics cards. New media and novel input/output devices have also extended the range. This transition in interaction speeds has promoted the development of different interfaces, such as virtual reality or interactive video. Many homes have accepted 'Broadband' as part and parcel of their everyday living and prefer this higher speed access to slower 'dial-up connections'. The rival service providers are now pushing the boundaries of home connection speeds beyond 8Mb – offering much greater service potentials and the user's wider freedom in what they transmit over the connection.

- The spread of systems with high functionality. Systems increasingly offer a larger number of functions associated with them which in itself will bring about changes in how the humans will adopt to use of the device. Many IT systems now come with onboard tutorials and help menus which obviate (or at least reduce) the need for traditional approaches such as paper manuals.
- Mass availability of enhanced graphics functionality. The graphics capabilities of most computers are such that they are now able to handle applications such as image processing, transformations, rendering, and interactive animation with reasonable ease. The functionality of these cards allows most screens to accommodate resolutions in the order of 1024 x 768 pixels.

With increasing fidelity of the images provided to us we need to consider the issue of visual impact. The human brain still receives significant quantities of information through the eyes. The broad matter of visual impacts includes:

- Mixed media and augmented reality. Most systems are capable of handling images, voice, sounds, video, text, and formatted data. The separation of consumer electronics (e.g., stereos, DVD's, VCR's, televisions) and computing has begun to disappear and the 'home centre' is available (albeit currently at relatively high prices). The future of such integrated devices is strong and a number of leading players in both fields (e.g. Phillips, Hewlett-Packard) are developing or have formed alliances with the intention of bringing integrated consumer products to the market as quickly as possible. Systems from Hewlett-Packard, Gateway and ZT Group have demonstrated the capabilities of devices that integrate TV, movies, music, slide shows and Web surfing in one.

Convergence between computing and entertainment is demonstrated by systems with a wide range of audio-visual connection sockets (jacks) and extra drives that make connecting to other components in the home relatively simple.

- Large, thin, novel displays. The days of the bulky Cathode Ray Tube (CRT) are clearly numbered: even Liquid Crystal Displays (LCD's) are running on borrowed time. New display technologies are maturing enabling very large displays and also displays that are thin, light weight, and have low power consumption. Users will see significant developments over the next few years, with ongoing development of products that use Organic Light Emitting Diode (OLED) displays. This technology will change the range of colour options available on current and future devices (gone are the days when they were mono-chromatic).

The current development of 'roll able' screens suggests there is limitless potential for portability of devices.

The effects on portability are beginning to show through, with many laptops/notebooks weighing less than 2Kg. The development of paper-like and pen-based (intuitive) computer interaction systems continues. These interfaces will be very different in feel from the desktop workstations many are familiar with. The development of new displays capable of delivering both 2D and 3D pictures to several viewers simultaneously attracts considerable research funding and their adoption on a range of screens/devices will bring new perspectives to the objects we view on a daily basis (http://www.seereal.com/). Augmented reality systems are spreading through a range of and have enormous potential for visualisation of buildings and other design aspects. Having considered these factors, we need to develop an understanding of and grasp the basics of, pervasive computing:

As Information and Communication Technology (ICT) continues its march towards full integration within our environments: everyday consumer items such as toys, milk cartons, and cars, along with workplaces and whole urban areas – interface and mesh with integrated processors, sensors, and actuators (linked via high-speed networks) and combine with new visualisation devices ranging from projections directly into the eye to large panorama displays, to form a seamless system which supports our needs and also the whims.

The goal of developers and researchers being the creation of a system that is pervasive and unobtrusive, embedded in the environment, completely connected, intuitive and portable, and available 24 hours a day seven days a week, 52 weeks of each and every year. Organizations developing pervasive computing systems see the underlying tenet as being akin to the provision of oxygen: they see a future where computing devices are ubiquitous, freely available and easily accessible (although whether the devices or services will be free or not is a matter for some debate).

These three previous factors then coalesce to affect and shape users as individuals, as members of groups or teams, and within specific areas of the work environment:

- Group interfaces. Interfaces are widely available which allow groups of people to communicate and coordinate simultaneously. Many meetings relating to overseas projects are 'virtual'. These virtual groups have influenced how organizations deliver their products and services: the nature of the organization and the division of labour has moved to accommodate new practices.

Ordinary users have the potential to tailor applications to suit their specific needs and will utilise this power to conjure up applications based around their understanding of their own domains.

Publicly available information is spreading at ever increasing speeds with novel delivery solution such as home banking and shopping becoming readily accepted practices. The rate of proliferation will accelerate with the introduction of higher-bandwidth interaction and the improvement in the quality of interfaces. IT systems are blending into the wider business environment and becoming much more intimately associated with the user's activities. Personal computers are becoming ever smaller and more personal which leads to developers having to be able to answer questions on the design of appropriate interfaces so that the user can operate them with the minimum discomfort and reduced stress.

3.3 HUMANS AND WORK STRESS

As stated previously, the focus of this book is about how IT has been implemented in a number of organisations. It can be argued that all organisations basically comprise two things: people and jobs. It doesn't matter if the organisation is profit or non-profit making; work (the job/s) still has to be undertaken by humans.

Given that those in employment will spend around a third of their life in the work environment, it is important to ensure that that environment is conducive to work. In an IT context we must consider a number of factors as diverse as social organization in the workplace, human information processing (including language, communication, and interaction) and ergonomics. All of these issues may work in isolation, or in combination, to bring about what is perceived as being a stressful working environment.

Employment, when viewed laterally, may be seen as social organization in the workplace. We must focus in on, and comprehend, humans as interacting social beings. Rightly then we must be concerned with the nature of work itself, and with the underlying premise that human systems and technical systems mutually adapt to each other and therefore must be considered holistically. Those tasked with implementing the IT system need to understand how small groups react as opposed to teams, and also how the individual reacts when faced with new or novel workplace issues. Work can be modelled and the models will in themselves influence the workflow, the degree of co-operation (and which activities this co-operation relates to) required amongst the participants and also the broader socio-technical system/s.

What has to be understood is whether the impact/s of the IT system is clear and its influence on the extant socio-technical systems and quality of work life and job satisfaction are well managed. This necessitates a solid appreciation of the likely user guidance required, the help techniques available and likely to be needed, the level of sophistication in the documentation provided and, how problems/errors will be handled. These of course depend to a large extent on how the human processes information.

3.3.1 Human information processing

In order that we can begin to grasp the impact of information on business it is important that we comprehend the characteristics of the human as a processor of information. A number of factors influence each of us as we process information and these include: cognition, memory, perception, motor skills, action and the underlying motivator. Discussion on a range of these areas is intended only to highlight their importance on how humans interact with, process and manage information, and IT systems.

3.3.2 Using sensory-motor skills

Sensory-motor skills form an important category of learning in the numerous tasks and occupations enveloped within today's complex businesses. Motor skills are normally classified as continuous (e.g. tracking), discrete, or procedural movements: the last category of skills being the more relevant to business IT applications (e.g. typing).

Long-term retention of motor skills depends upon regular practice; however, continuous responses show less forgetting in the absence of practice than discrete or procedural skills (consider how familiar you are with a specific task inside a piece of software you haven't used for some time e.g. constructing a pivot table in a spreadsheet). There can be no doubt that learning and retention of sensory-motor skills improves with both the quantity and quality of feedback received under the training regime. There is solid argument for the view that mental rehearsal, especially involving imagery, facilitates improved performance. Perhaps because it allows additional memory processing related to physical tasks (e.g., the formation of schemata) or because it helps in maintaining arousal or motivation centred on a specific activity.

A schemata is a mental mapping and structure we use to organize and simplify our knowledge of the world that surrounds us. We hold schemata about ourselves, other people, IT systems, software, a range of devices, food, and in fact almost everything that we know or come into contact with. Schemata can be related to one another; often in a clearly defined hierarchy e.g. I am British, Scottish, an Academic, male, a human.

Schemata influence and affect all that we do and say i.e. what we notice, how we interpret or deduce things and how we arrive at a range of decisions, and ultimately act. They operate in a fashion similar to a filter; they mitigate some things and accentuate others. Humans use them to classify things, think of how you may have 'pigeon-holed' a person – that is a schemata being applied. Schemata also help us in forecasting and predicting what will happen: we even remember and recall things via schemata, using them to 'encode' the very memories we hold. Schemata are often shared within cultures, thus allowing short-cut communications which are nonetheless fully comprehensible. In essence every word we use can be viewed as schemata: the instant you read/see it you receive, and subconsciously unravel, a package of additional inferred information.

Numerous works have fed directly into theories of learning and skill acquisition, an important aspect when we sit a human down in front of new piece of 'kit' or present them with new software. It is not enough to simply leave them to fend for themselves. We have to grasp how they perceive this new challenge, if it is even perceived as a challenge.

Models of human cognitive architecture are plentiful and they aid in developing software that mimics or at very least complements the operations of the brain. The memory aspect of the brain has been the focus of considerable research (see for example the writings of Uri Fidelman) and is an area with considerable development potential. Memory has yet to reveal all its secrets and holds many frustrations for researchers: we try to remember all sorts of things but often forget them: how often have you heard someone say 'it's on the tip of my tongue', but they can't actually recall the answer.

Fidelman (2002) when writing on temporal and simultaneous processing in the brain, and a possible cellular basis of cognition, argued that two factors may underpin all cognitive functions: the first being the firing rate of neurons; the second being the likelihood of neural transmission errors. The net effect being a moderation of both the cognitive style and the efficiency of the neural processing. The neural processing drawing on the memory of the individual. Comprehension, or attempts at comprehending, memory continue unabated: much attention has been given to the number and types of branches and descriptive models, for example:

- The multi-store model: where memory is construed as a series of stores.
- Sensory memory: buffering the most recent of data.
- Short-term memory: which suggest we all have a limited number of slots.
- Long-term memory: which is capable of holding an individual's entire life?
- Storing (encoding) memories.
- Enhancing memory: approaches that are designed to improve recall.
- Encoding specificity principle: where memory is related to a specific situation or context.

Phenomena and theories of memory have been considered by numerous writers e.g. Furnham, Weiner, and what has emerged from these writings is a view that there is a clear set of branches to the conceptualisation and understanding of 'memory' and each of these branches are forthright in their importance of their area. Again, it is not the intention to cover all branches rather a brief discussion on them will provide a flavour of the area and set the scene.

Primacy Effect: essentially the argument is that we remember what happened first. Given a list of items to remember, most individuals will tend to remember the first few things more than those things in the middle or at the end. The primacy effect has most effect during a repeated message when there is insignificant delay between the delivery of the messages.

One reason that the Primacy effect works is that the individual is more likely to commence by paying attention, then drift off when the subject gets boring or their internal processing of the data assumes prepotence. The limitations of the individual's memory also has an effect, and we can miss middle items as we continue to rehearse and process the initial items.

Implementing IT means that we also have to appreciate phenomena and theories of perception since they relate directly to attention and vigilance. The following diagrams (Figures 3.3, 3.4 and 3.5) are intended to challenge your perception (answer the question located next to each diagram). The work involved for you in each of the diagrams is fairly simple, in that you have to either state what you see (or think you see) or else state which proportion you think is greater. They are diagrams designed to challenge: accept them for what they are.

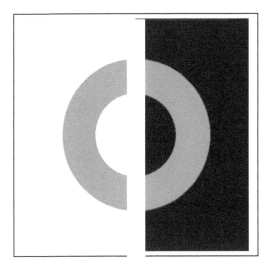

Figure 3.3 Which side of the circle is bigger? Answer =?

Having pondered on the circle and its proportions, we presume you got the answer correct at the first attempt, let's keep the perception problem going and have you consider the diagrams in Figures 3.4 and 3.5 In Figure 3.5, the trick is to blink once or twice, rapidly, then look at the figure and answer the question set. Figure 3.4 quite simply requires you to answer what happens and yet the answer may not be that easily arrived at.

Figure 3.4 Stare at the centre of the diagram: what happens? Answer =?

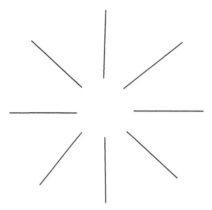

Figure 3.5 What is at the centre of this diagram? Answer =?

Users' conceptual models and models of human action rely on the fact that human diversity is a constant force which ensures innovation in approach – no two people will be identical and therefore IT systems designers must take account of this diversity in approach. When we include disabled individuals (with whatever form of disability) into the broad equation then the system developer's challenge is to ensure all designs accommodate a wide range of human actions (some voluntary and some involuntary e.g. consider a sufferer of Tourette's syndrome).

An individual's voluntary actions and what drives one person and not another have long been the focus of many writers on theories of motivation. Many models have been proposed e.g. Expectancy, Hierarchy of Needs, etc. What they all have in common is the ability for at least part of the theory to be taken and applied to humans at work in a variety of situations. Which one aspect is best is contingent upon a number of issues including: the individual, the situation, the management style, and the task. Conceptual models have been developed and used with varying degrees of success.

Given that some drivers of how humans process information and use it in delivering targets are now understood, it allows us to move on and consider the role of language, communication and interaction in an IT setting. Language serves as a communication and interface medium. Various aspects of language which are used to engage with other humans e.g. the syntax, semantics, pragmatics, are actively seen as mechanisms by which the individual can be allied to the IT system. For instance turn-taking is a pragmatic phenomena which is routinely accepted when communicating via electronic medium (if everyone tried to communicate at the same time, at least part of the message would be lost).

Specialist languages such as graphical interaction, query, command, and editors have enabled most competent IT users to become adept at conversing in new languages. Indeed many individuals have voluntarily engaged in developing some degree of expertise in these specialised languages. Couple these skills with the correct working environment and we have all the ingredients in place for ensuring that we have highly productive IT workers. The ergonomics of the workplace often being an issue which is overlooked and yet plays a vital part in successful IT implementation.

3.3.3 Ergonomics

Most architectural design these days relies on a wealth of anthropometric data and an understanding of the physiological characteristics of people in order to develop a solid understanding of humans in the workplace. The Architects' Journal publish a solid reference book on anthropometric data and its use in the built environment. This grasp of the relationship to workspace and environmental parameters suggests that we include consideration of factors such as:

- Human anthropometry in relation to workspace design – space and facilities for the tall are not the same as those required for the shorter individual.
- Arrangement of displays and controls requires to be analysed in order that solid workflows are established which are conducive to sustained output. Human cognitive and sensory limits have to be borne in mind – overload leads to problems with 'burn-out'.
- Sensory and perceptual effects of display technologies, including their legibility and display design have significant effects on the individual. Specific legislation on Display Screens, etc. exists to ensure that safe working limits are not exceeded.
- Furniture and lighting design are important in order that the human is not fatigued. The use of IT normally relies on a range of surfaces being utilised and these surfaces must be conducive to sustained effort.
- The surrounding temperature and environmental noise are issues which must be managed. Whilst there is legislation which sets a minimum working temperature, often there is no maximum temperature advised which can lead to fatigue problems for the worker.

There is an inherent assumption in this book that design of the workplace includes design for the inclusion of the disabled. The range of disabilities means that those designing for inclusion must often think laterally i.e. the disability may not be an obvious physical one. If all of these cognitive and workplace environment factors are considered carefully and incorporated into design of the workplace we can go some way towards reducing its effect as a stressor. However there are other stressors which must be accounted for.

3.4 STRESS AND PERSONAL ISSUES.

Stress can be defined as a by-product of demands made on the individual which they feel they are unable to adequately cope with i.e. when an individual feels their inherent capabilities and resources are unable to meet the demands placed upon them by a given situation (in this case the workplace demands).

The last 2 decades has seen a consistent surge in the recognition of job stress as a significant occupational health problem and the demand for measures which seek to manage the issue in a responsible, professional, manner. The much reported case of Walker versus Northumberland County Council (AELR, 1995) established the UK precedent for future litigation and case structures, based on job stress.

Job stress is invariable brought about by a combination of forces acting on the individual i.e. its source is multi-variate. The multi-variate model of job stress shows that the potential causes of occupational stress are numerous and varied: broadly speaking there are five categories of job stressors which are modelled in Figure 3.5 below.

French and Caplan (1973) considered both quantitative and qualitative aspects of situations as primary stress factors. To these specific factors, other influences such as individual differences and non-work factors may be added to provide a composite model as shown in Figure 3.6.

The influence of single, or combined, factor/s may induce job stress and result in what is perceived as 'losses' to both the organisation and the individual. It may be that there is an optimum level of stress at which the individual works effectively i.e. there is a plus side to stress. In the workplace, the IT manager is tasked with finding the impact of IT systems on this optimum level. The other side of the coin is that the absence of stress may lead to what is termed 'rust-out': where the individual feels under utilised. Whilst too much stress may lead to the inevitable 'burn-out', either of these situations may cause the individual to suffer distress.

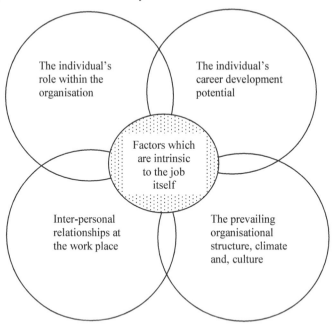

Figure 3.6 The interacting job-stressors

Stress then can be more widely perceived as the rate of wear and tear caused by the combination of 'work-life' and 'normal' life (Figure 3.7 illustrates the interaction of these forces). Job stress has adverse effects which have been clearly highlighted in a number of research publications; the effects including: poor job performance, high levels of absenteeism, discontent or low morale and high labour turnover, this last feature resulting from the loss of 'wanted' employees, the removal of 'unwanted' ones and retirement from work through ill-health of those who on the face of it were contributors to the organisation (a serious problem when the construction industry is in 'boom' mode).

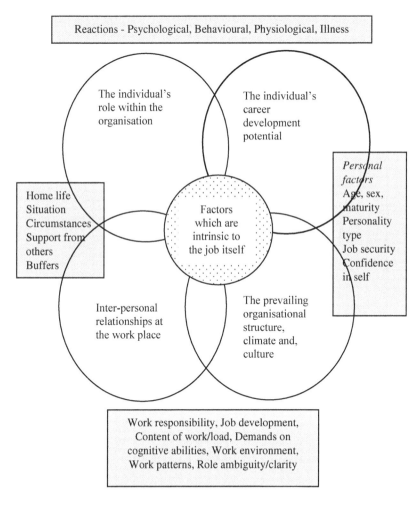

Figure 3.7 The interacting stressors

The costs attributable to stress have been considered in a range of industries and have been well discussed and documented over the years. Figure 3.7 shows a composite model of a number of the likely cost areas. It becomes clear from Figure 3.7 that the composite cost, at various points in time, is comprised of several key elements which may be manipulated to reduce the summative effect.

Analyses of absenteeism data available suggested that as much as 40% of absenteeism could be attributed to mental or emotional problems; and that the incidence of these is increasing (even though the various health and Safety at Work Acts in the UK have placed a specific duty on the employer to ensure the mental wellbeing of the employee). Litigation then is an additional element which must be factored into, and added to, the broader cost equation. Stress cases which have gone through the courts have lead to awards which have ramifications for those managing human resources and will no doubt add substantial sums to every employer's cost structures.

Efficiency measures being implemented by many organisations may well add to an employee's stress in terms of anxiety and uncertainty of employment. In order to redress the balance, the organisation may require considering and implementing, a strategic approach to stress management.

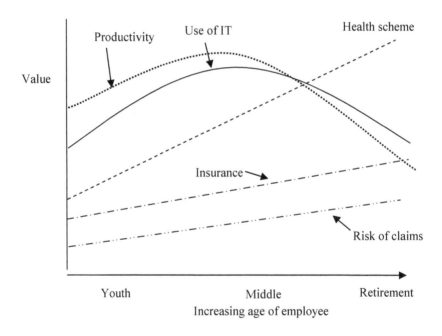

Figure 3.8 Impact on the overall cost of employing an individual

3.4.1 Options open to those managing

Management faced with employees who see themselves as operating within what are perceived as stressful situations should have a range of strategies available for rapid implementation. These approaches may encompass one or more of the following:

- Enhancing positive coping strategies; and,
- Implementing stress management and counselling programs; and,
- Psychometric tests applied to individuals.

A robust stress management regime must be put in place: court cases are traumatic for all involved and should be avoided. There is no doubt that organisational stress is a sensitive issue, as such it calls for close examination of work activities and the people involved and the facilities they utilise i.e. the IT.

There are benefits to be gained from 'well' employees who operate in a sound IT environment. A more efficient employee may produce better work and sustain that level of performance for a significant period of time. Part and parcel of the creation of this sound IT environment entails supportive training.

3.4.2 Training for IT

Skills and training for IT can be both generic and specific. Generic in the sense that the individuals undertake training in common software or the application of this software e.g. spreadsheets and their use in financial modelling: specific in that many organisations will take such software and develop bespoke applications around its core. The ECDL (European Computer Driving Licence) is widely accepted and taken as being a fundamental mark competence in IT with its coverage of a number of key areas. The ECDL is an internationally recognised qualification (there is also an ICDL) enabling numerous people to demonstrate their acquired competence in computer skills. It is rapidly achieving the status of 'required standard' within both large and small employers alike. Organisations such as the education sector, the NHS, the MOD, the Bank of England and a host of financial institutions actively encourage its uptake.

The tests for each of the ECDL modules can be taken at more than 2,500, widely dispersed, approved UK test centres. To obtain the qualification it is necessary to register with the British Computer Society (the UK accrediting body) before taking the tests.

In excess of 2 million people have finished or are part-way through studying for their ECDL. It is widely seen as the leading certification standard for basic computer (IT) and Internet competence. It demonstrates to potential employers that the individual has attained a recognised, relevant qualification and also helps the individual in using their computer with increased confidence and competence.

How ECDL works, well the ECDL syllabus is designed to cover the key concepts of computing, its practical applications and their use in the workplace and society. It is broken down into seven modules, each of which must be passed before the ECDL certificate is awarded.

Once a candidate is registered at an accredited Test Centre, a logbook listing all 7 modules will be issued. The modules can be taken in any order and over any period of time (normally up to three years) offering the individual a high degree of flexibility. When all 7 modules have been successfully passed, the logbook is exchanged for a certificate and ECDL 'licence' card. Benefits derived from undertaking the ECDL include: raising the individual's level of competence in IT and computing skills, improvements in work productivity, and attainment of a worldwide recognised qualification. A brief synopsis of each of the ECDL modules follows and provides a glimpse of the broader constituent elements of the qualification:

Module 1: Basic Concepts of IT - Discover how IT systems are found in everyday situations, including health, security and legal issues.
Expanded syllabus
Requiring the candidate to have an understanding of a number of the main concepts of IT, at a general level. The candidate will be required to understand the make-up of a Personal Computer (PC) in terms of hardware and software and to understand a number of the concepts of Information Technology (IT) e.g. data storage and memory. The candidate shall also understand how information networks are used within computing and be aware of the uses of computer-based software applications in everyday life. The candidate shall appreciate health and safety issues as well as some environmental factors involved in using computers. The candidate shall be aware of some of the important security and legal issues associated with computers.

Module 2: Using the Computer and Managing Files - Understand the basic functions of a personal computer and its operating system.
Expanded syllabus
Requiring the candidate to demonstrate knowledge and competence in using the common functions of a PC and its operating system. The candidate shall be able to adjust main settings, use the built-in help features and deal with a non-responding application. They shall be able to operate effectively within the desktop environment and work with desktop icons and windows. They shall be able to manage and organise files and directories/folders and know how to duplicate, move and delete files and directories/folders, and compress and extract files. The candidate shall also understand what a computer virus is and be able to use virus scanning software. They shall demonstrate the ability to use simple editing tools and print management facilities available within the operating system (OS).

Module 3: Word Processing - Learn the basics of word processing. Progress to more advanced areas of Word including tables, images and mail merge.
Expanded syllabus
Requires that they demonstrate the ability to use a word processing application on a computer. Candidates shall be able to accomplish everyday tasks associated with creating, formatting and finishing small sized word processing document ready for distribution. They shall be able to duplicate and move text within and between documents. Candidates shall demonstrate competence in using some of the features associated with word processing applications such as creating standard tables, using pictures and images within a document, and using mail merge tools.

Module 4: Spreadsheets - Understand the basics of spreadsheets before moving on to more advanced areas including importing objects and creating graphs and charts.

Expanded syllabus

Requires candidates to understand the concept of spreadsheets and to demonstrate the ability to use a spreadsheet application on a computer. Candidates shall understand and be able to accomplish basic operations associated with developing, formatting, modifying and using a spreadsheet of limited scope ready for distribution. They shall also be able to generate and apply standard mathematical and logical formulas using standard formulas and functions.

Module 5: Database - Learn how to design and plan a simple database. Understand how to retrieve information using available query, select and sort tools.

Expanded syllabus

Requires candidates to understand some of the main concepts of databases and demonstrate the ability to use a database on a computer. Candidates shall be able to create and modify tables, queries, forms and reports, and prepare outputs ready for distribution. Candidates shall be able to relate tables and to retrieve and manipulate information from a database by using query and sort tools available in the package.

Module 6: Presentations - Design simple presentations using basic slide show effects and the use of graphics and charts. Create a range of different presentations suited to different target audiences or situations.

Expanded syllabus

Requires candidates to demonstrate competence in using PC based presentation tools. Candidates shall be able to accomplish tasks such as creating, formatting, modifying and preparing presentations using different slide layouts for display and printed distribution. They shall also be able to duplicate and move text, pictures, images and charts within and between presentations. Candidates shall demonstrate the ability to accomplish common operations with images, charts and drawn objects and to use various slide show effects.

Module 7: Information and Communication - Includes tutorials on effective use of the World Wide Web, including web searching and web browsing basics. Communication introduces you to e-mails and covers sending and receiving attachments and organising e-mails into folders.

Expanded syllabus

Two sections - firstly, Information requires candidates to understand some of the concepts and terms associated with using the Internet and to appreciate some of the security issues. They shall also be able to accomplish common web search/web browsing tasks using available search engines. They shall be able to bookmark web sites, and to print web pages/search outputs. Candidates shall also be able to navigate within and use web-based forms. The second section, Communication, requires candidates to understand some of the concepts of electronic mail (e-mail) together with an appreciation of some of the security issues associated with using e-mail. They shall also demonstrate the ability to use e-mail software to send and receive messages, and to attach files to mail messages and file mails.

Whilst the ECDL syllabus normally contains material covering applications such as MS Office 2000, XP and 2003 and revolves around: Word Processing, Spreadsheets, Databases and, Presentations, there may be an advanced/additional syllabus which extends these areas and offers specialisms such as CAD and CAFM. ECDL is also available on the Internet and there are a number of online service ECDL providers who offer a host of features such as:

- Interactive tell-show-try videos with full audio support.
- Practice exercises and assessments.
- On Line support materials.
- Progress tracking system.
- Easy to navigate menu system.
- Fully supported Distance Learning Option (subject to availability).

Other training to fit with business needs and the needs of the individual may be required. In order to ascertain the current position relative to the needs of the individual, and management should undertake and complete some form of Key IT Skills Self-assessment Audit. The findings form the audit will help determine key areas and inform the later choices.

3.5 SUMMARY

This chapter has shown that those challenged with implementing any form of IT must be aware of the 'softer' issues surrounding the IT. Stress is a debilitating effect of work (and other) pressures which can lead to severe effects for the individual and the organisation. Sound management of the IT implementation demands sound management of the human resources involved.

Humans interact with the computer and IT systems and because of this interaction a solid appreciation of HCI as a concept provides those involved in IT implementation with a grasp of the many factors influencing not only the design of the hardware and software, but also the environment that surrounds the IT operator.

Part and parcel of the 'business environment' are the business processes actually undertaken within the organisation, at a range of levels. As mentioned earlier, all organisations can be viewed as basically comprising two things: people and the jobs. The IT systems and solutions implemented within the business have to be selected on a basis that has taken account of what has to be done and why in a particular fashion, who will do it and how they will carry out the tasks and, when and where the work will actually be done. It is this concern for the processes/tasks undertaken and how IT utilisation brings about improvements that forms the focus for the next chapter. Before moving on, consider the questions and points to ponder.

3.6 QUESTIONS TO PONDER

- Each and every organisation still relies on humans to undertake a range of jobs: why has there been such a reluctance to increase the range of functions undertaken by IT based systems e.g. automatic interim valuation generation via an integrated supply chain?

- The design of the workplace can be a significant influence on the health and well-being of the employee: questions arise as to why are so few people aware of the contents of the various pieces of legislation e.g. the Health and Safety Display Screen Equipment Regulations 1992.
- The numerous writers on motivation and behaviour in the workplace would have us believe that people like challenge: novel IT systems are a challenge and yet many individuals fear this challenge. How do we overcome the fear element?
- The underlying demographics of the industry are changing which suggests that IT literate staff will become the norm, rather than an exception. What level of investment in training should we commit to the existing workforce?

CHAPTER 4

Contemporary business processes: how IT improves these processes

This chapter will consider:

- The need for collaboration within the construction industry.
- Information management within the construction industry.
- Creation of information and its destination.
- The flow of data and transfer of documentation.
- Communication within the construction industry.
- Inadequate information through non-integration: conflict and claims.
- The use of documentation within arbitration.
- Decision making within the construction industry.
- A summary.
- Questions to ponder.

4.1 INTRODUCTION

The UK Construction Industry has long been seen as an industry, which is highly fragmented and non-collaborative. This highly fragmented nature of the Industry coupled together with the risk and uncertainty involved has meant that clients and contractors alike actively look for ways of improving overall quality and performance. The current nature and scope of the industry however means many processes are being replicated resulting in waste and inefficiencies amongst project partners. In order for the industry to improve its performance the industry needs to change its prevailing culture towards a culture preached by both the Latham (1994) and Egan (1998) reports which support continuous improvement by adopting collaborative working practises between clients and contractors. This collaborative culture should also allow for the possibility of information to be shared between projects and teams and across organisational boundaries. Team syntegrity will also support the construction sector in adopting new processes, which at the end of the day improve productivity and quality.

Those who become involved in collaboration projects should experience no disadvantages when adopting current collaborative working practises although some members of the collaboration team do have to work at it harder than others in order for the collaboration to succeed (Yeomans et al., 2006). The many benefits of collaboration include elimination of problems at design stage, installing confidence that will help to eliminate problems, creation of a better working environment, greater co-ordination and an increase in project efficiency. For collaboration to succeed however other factors have to be considered for the successful, implementation of collaborative working including: (Yeomans et al., 2006).

- Everyone must be committed to the adoption of collaborative practises.
- Practises must be used alongside existing procurement systems which are not specifically tailored for collaborative working.
- There are no easy to use systems (models and tools) for capturing and reporting the benefits and this can slow the uptake of collaboration tools.

Modern information technology (IT) has now become a crucial component within construction resource management and the use of IT within the construction industry is now becoming more widespread as awareness of its benefits increases. Recent IT adoption within the construction sector is beginning to lead to construction disciplines becoming more open and honest with each other by sharing project information electronically although the adoption of IT and in particular document management systems within the construction industry has not been as rapid as other industries such as the manufacturing industry. Construction organisations are only now beginning to understand how crucial the roles of IT systems are in the exchange of project information. Currently, the use of IT in construction to manage and control project information has been the automation of tasks and processes that were previously carried out manually. Over the past few years however notable developments and the implementation of such IT systems have been recorded in this field due to the pressures being imposed by clients on contractors to improve productivity and obtain value for money within construction projects. As early as 1995, the Department of the Environment (DoE) developed an IT strategy that consisted of three main elements, which were:

1. Encouraging improved sharing of information through the use of integrated project databases. In other words 'IT'.
2. Developing an industry wide knowledge base that facilitates the sharing of information and promotes team syntegrity.
3. The use of IT to improve basic project processes.

Whilst other industries have utilised the application of IT, the construction industry has remained a cost avoidance industry with a relative declining product quality. The construction industry therefore requires integrated collaborative systems that facilitate this task and the development of these systems offers the potential to radically change the structure of the industry and the way information is currently passed between organisations. Furthermore, demands being made by the client to improve the overall quality of the end product have lead to increases in productivity, which helps to enhance the overall image of the industry. The end result is a complete collaborative, integrated project life cycle that fuels the industry and consistently ensures fewer errors are made and which promotes integration amongst the industry's distinct separate professions.

An analysis of current working methods with regards to information management will demonstrate that standardised IT systems are the way forward if documentation is to be controlled, structured and readily available to all construction professionals at the touch of a keyboard. Controlling the creation, distribution and administration of all project correspondence in a systematic manner supports integration between a range of construction organisations, both geographically and temporally displaced.

4.2 INFORMATION MANAGEMENT

Within the construction industry there is a growing demand for integrated information systems that allow different participants involved in the project to share common project information. Construction projects themselves have significant embedded information, and generate considerable quantities of real time information prior to, and during execution. Management of the construction project and in particular management of information and documentation therefore needs to be structured in a logical order using a system or a combination of systems that ensures the project participants have instantaneous access to all project information. These systems facilitate data and information input, access, retrieval and functional bi-directional management.

The onus of storing and managing project information received has usually fallen to each individual member of the project team. Dramatic advances and vastly improved IT Systems facilitate synergistic standardised filing and information control, which obviates the reliance on each individual or organisation to undertake this task.

The issue and collection of this information on construction projects however is still largely a traditional process mainly carried out using pen and paper. Many information management tasks that could be automated and re-engineered are still paper based. Given that many tasks and projects are geographically distributed the exchange of information is slow and arduous resulting in delays and errors. The lack of an integrated information management system will lead to the information becoming lost because it is not stored in the appropriate format which effectively means others who may wish to use the information are deprived of the chance to do so. However, many construction professionals are becoming increasingly IT literate although the vast majority of construction professionals still insist on using these traditional communication methods to collect and transfer information.

What is required is a combination of IT technologies that will allow the construction organisation to continue using its current processes for collecting information and documentation. The implementation of document/information management systems and associated mobile technologies will allow construction organisations to become IT orientated whilst at the same time allowing the remotely based construction operative the opportunity to collect and transfer project information electronically whilst at the same time allowing them to work in a manner which they are familiar with i.e. pen and paper.

In addition projects are becoming more complex which means that the amount of information contained within grows at an extraordinary rate. Construction organisations need to recognise the importance of information management with regards to the creation, gathering and storing of information. The use of IT systems would allow all project stakeholders to make smarter requests for drawing detail's, requests for instructions, requests for project information and general administration procedures. The potential benefits of collaboration are numerous and include cost savings, time-savings, productivity increases, promoting partnerships and increasing harmony amongst project members. These benefits will hopefully lead to construction professionals becoming more open with each other by sharing project information electronically by using an information management tool or central data repository.

The sharing of information is essential to the running of any project and the adoption of an IT system will aid the project process. The transferring and sharing and responsibility for information within the project environment can also be dependent upon who actually has access to the information and who is allowed to create and manage project information although the actual management of information on a construction project will usually be the responsibility of the main contractor who will report and pass on information to the client. If this structure changes however, conflict and claims situations will arise as decisions will be made on the basis of imperfect information received. This in turn leads to delay and disruption, which can cause the project to fall behind schedule.

The main objective of information management in construction is to achieve coherent management and electronic sharing of information during construction projects. Construction projects therefore require Information Management Systems (IMS) that facilitate this task, which force construction disciplines to collaborate, co-ordinate and co-operate with others which will ultimately mean more relaxed and smoother project processes. The solution therefore is an IT system or a series of IT tools that will aid the construction process and manage the creation and distribution of project information. As the culture of the construction industry changes and embraces technological solutions, the aim is to make IT the 'tried' and 'tested' method of doing business.

The use of project databases and the Internet will encourage the free transfer of information between all parties within the contract before finally processing the finished product through the necessary formal channels. As IT adoption increases amongst the industry's leading players, the use of integrated information and document management systems will become embedded within the overall construction delivery process. First of all however, we must begin to understand how documentation and information is currently created and transferred within construction projects. This will involve an examination of the processes the construction industry uses for capturing information. We must understand the need for IT and build up the case for its adoption and implementation. Dealing with information in real time and providing the necessary and correct information through the adoption of an information management system is essential if construction projects are to become better more flexible collaborative organisations.

4.2.1 Information in a construction project context

A construction project is a highly complex collaboration activity involving many different bodies and organisations such as designers, clients, consultants and contractors. Cross discipline communication between these distinct professions is very often problematic and a major contributory factor to poor project performance is the lack of integration and co-ordination between the industry's distinct professions.

With a growing emphasis being placed on meeting client needs and improving project performance, the construction industry is at last beginning to use fully integrated project teams and project processes where project participants have instantaneous access to all project information. The fluctuating nature of construction contracts and the increased vulnerability of project personnel have often meant that project information is not transferred, stored and disseminated the way it is intended to be. Over the duration of a project, individuals and organisations will accumulate huge amounts of real time information. An understanding of how, why and when information is created and disseminated will allow an organisation to identify current process bottleneck areas. This in turn will lead to more efficient information management processes.

4.2.2 Information – creators and receivers

As IT usage increases within the construction industry it is hoped that the quality of documents, communication channels, speed of work and simpler access to project information will become common place. Many major construction clients are now demanding computer integrated construction processes. The concept is for all project participants to develop the ability to communicate and transfer project data electronically.

Traditionally the co-ordination of the numerous parties involved in a project has been a daunting and chaotic experience and has mainly been the responsibility of the main contractor. The end result as seen in Figure 4.1 is a communication process at project level which is unstructured and chaotic. Parties to the construction project still rely on personal contact, telephone and paper to manage information created within the project environment. The use of paper as a form of communication is still undoubtedly the main medium and although the use of IT is increasing, the electronic exchange of information at the construction stage is still in its infancy.

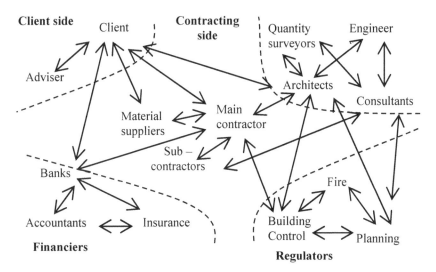

Figure 4.1 Traditional chaos in communication within the construction industry

Bowden (2005) indicates that site based personnel are both senders and receivers of paper based documentation which can be grouped into different document types revealing the most common tasks as following:

- Completing data collection forms (25%)
- Dealing with general correspondence (18%)
- Viewing and receiving drawings (13%)
- Reading, interpreting and writing specifications (6%)

The research went further and also found the types of documentation that site based personnel would like to have access to at the point of their work. The most common document types that site based personnel would like access to in the field are (in ranked order):

- Drawings (24%)
- Data collection forms (12%)
- Correspondence (8%)
- Progress information (7%)
- Specifications (7%)

Issues that can have a profound affect on communication tend to be centred on dissemination, transfer, human attitudes and protocols. However, as the construction industry begins to support integrated collaborative ways of working the change from the traditional way of communicating to a new way of communicating electronically will take time to complete but will result in a communication process that is structured and co-ordinated throughout the construction supply chain.

4.2.3 Data flow and transfer of documentation in the construction industry

The life cycle approach to construction contracts has resulted in the construction industry being challenged to capture, for future use, the information and documentation generated. This in turn has lead to the more effective use of IT. Crucial to the running of any major construction project is the movement of project information amongst the industry's professions, all of whom however have conflicting priorities and objectives both individually and collectively. This problem which in effect relates to communication is exacerbated because each of the differing professions all have or use their unique information and documentation processes to undertake tasks. However, to undertake these tasks effectively they have become heavily reliant upon information that has to be supplied by other professions to satisfy their own contractual responsibilities.

The sheer number of contractors, sub-contractors and industry professions involved in larger projects also causes great difficulty for the project management team who try to 'manage' the communication channels between them. The construction stage of a project is heavily dependent upon the flow of data, the sharing and the exchange of data at the construction stage although traditionally the flow of data is biased towards traditional methods of communication (paper). These unique professions and the industry's fragmented culture towards information and documentation would suggest that the adoption of IT will aid collaboration and cross discipline communication of the numerous players involved within a project. These inefficient traditional methods of communication have however become a barrier to the uptake of IT within the industry. In effect however the industry's unique professions rarely acknowledge the needs of each other and information/documentation that is provided is rarely available in a suitable format and in some cases the information is so incompatible the next step within the information process is to totally re-construct what has been provided.

Who controls project information can often depend on who has created it. The controller of information within the construction industry could be the architect, contractor or client although this can depend upon the type of contract and procurement route adopted for the project. For example within the traditional route of procurement, it is the contractor who takes responsibility for the works/construction section and in reality it is the contractor whom which most communication and information will be channelled through although to some extent the client has some influence. Within the construction management route of procurement the transfer of information is not so clear-cut. Depending upon the nature of the project, overall responsibility for the project may shift during the different phases of the project. This implied flexibility of the management structure is essential if the contract is to be successful although communication breakdown will occur if there is a conflict of interests between contractual parties.

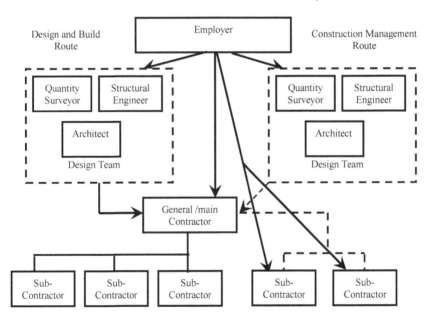

Figure 4.2 Construction management and traditional procurement routes: comparisons in the links between them

Figure 4.2 illustrates that the client is the only party to have formal contractual links with other parties within the contract. The client operating within the construction management route of procurement also has the right to communicate directly with the sub contractors unlike the design and build route where the client has to communicate through others. The client is therefore in a very strong position to try and influence others to adopt the use of IT that will force integration, aid the transfer of project information and increase communication between all parties of the contract. If the flow, sharing of and transfer of information is to become automated then some of the communication boundaries that currently exist within the industry need to be broken down. The sharing and transfer of compatible information needs to become commonplace which will create a culture where information is freely exchanged in a disciplined manner. New communication methods however depend upon trust and mutually beneficial stable partnerships. It is highly likely that the use of IT systems would offer great possibilities for the efficient conduct of conflict and claims situations that arise as a result of the poor flow or project information.

4.2.4 Communication in construction

The construction industry needs to communicate better not only with the outside world but also with its distinct internal professions. Communication can be seen as a professional practice and it is therefore not surprising that the majority of people within the construction project who communicate most often are the distinct professions such as the contractor, the client and the architect. The construction industry is an industry highly dependent upon information and heavily based upon traditional means of communication such as face to face meetings, phone calls and the exchange of drawings and associated paper documents. The amount of information generated and transferred during construction projects can be enormous even for the lower valued and less complex projects. The construction project therefore depends heavily upon the timely transfer of information because it forms the basis on which decisions are made by the distinct professions/professionals within the industry. The use of IT has provided the construction industry with advantages in terms of creation, consistency, transfer and access to project documentation and document communication can therefore be seen as the key factor in the overall success of a construction project.

Paper based communication remains the cornerstone of construction communication and although many organisations now use electronic methods such as e-mail to transfer documentation it is the hard copy that is invariably used as the permanent record of information. The traditional method of storage and creation is also inherently time consuming and currently results in a situation where paper based information becomes lost and is not easily retained and the construction industry is therefore faced with the on-going challenge of improving its current communication methods in order to become more information focused.

The highly differing professions and their multi disciplinary skills often limit the scope of communication amongst them. To support and improve communication in the construction industry many alliances and partnering initiatives exist which, improve the overall communication and interaction levels within the construction environment. Due to the distinct nature of the industry many project teams Part Company when the project ends and therefore the communication channels that have been formed are completely dissolved. On Private Finance Projects (PFI) however the communication channels and links tend to be strong due to the long life cycle of the contract where documentation is shared and transferred openly as part of the open but structured communication network.

Construction organisations must form effective communication links in order to realise the benefits of partnerships and alliances. Organisations that rely on co-operation and trust have been found to obtain lower costs for as long as the network is maintained (Cheng et al., 2001). It is critical to the success of a construction project that the distinct parties to the project are integrated using communication methods that allow for the sharing and exchange of resources amongst its members.

Using collaboration tools such as IT brings together geographically dispersed parties of the project and improves the overall communication network within the project environment. These mechanisms are well established and allow for communication within construction to be completed electronically.

4.2.5 The effects of inadequate information: non-integration

Conflict, claims and disputes in construction projects has spawned many works by authors seeking to identify the common sources of disputes within the construction environment (Kumaraswamy, 1997). Some would argue that the construction industry is inherently burdened with conflict and disputes as a result of increasingly complex projects demanded by clients. Within construction projects delays are almost inevitable, since we are unable to control all forces acting on the project e.g. the weather, and as a result of this, conflict and claims situations will arise (Alkass et al., 1995). The poor communication processes within the construction project environment can also add to this complexity, and can often be blamed for causing frequent delays to the project and generating conflict between the project partners e.g. through poorly made decisions. A construction project begins as soon as the first negotiations or network communications have taken place. A great problem with many projects and construction organisations is inherently poor, unstructured record keeping and document control throughout the project life. However the use of information/document management systems can be called upon to provide structured document based evidence that can be used to settle disputes although many systems both IT and paper based are not designed to support conflict situations because of underlying incompatibility issues.

The circumstances that lead to claims have to be dealt with, but in today's more open construction industry there are new approaches available to aid decision making and speed up the dispute resolution process. Disputes between construction parties are of great concern, although the effective management of conflict can be achieved but only if it is based upon reliable evidence or the correct information (Fenn et al., 1997). Conflict arising as a result of communication can be traced back to inadequate, inaccurate, inappropriate, inconsistent and most importantly late information. Kumaraswamy (1997) has identified many of the root and proximate causes of conflict, claims and disputes. Some of these root and proximate causes can be listed as follows:

Root causes
- Unrealistic time, cost and quality targets by (clients).
- Adversarial culture of construction industry organisations.
- Inappropriate contract/procurement type.
- Lack of competence and professionalism.
- Lack of integration amongst project parties.
- Lack of adequate project information.
- Lack of information from the client.

Proximate causes
- Little integration in parties.
- Inappropriate forms of contract.
- Inadequate contract documentation.
- Inadequate contract administration.
- Inadequate design/tender information.
- Slow response from client.
- Poor communication (Kumaraswamy, 1997).

These proximate causes escalate and generate themselves or through interactions conflict, claims and disputes as displayed in Figure 4.3.

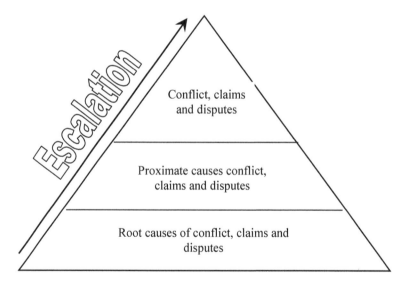

Figure 4.3 Conflict and claims escalation

This conflict can occur despite the advancements made in document management systems and information management techniques. Precise project information is important when trying to settle conflict events, although collecting and structuring the mountains of paper based documents created on most construction projects can be a tedious task made even harder because the information itself has been received from ad-hoc sources who each have differing information and documentation transfer procedures. Structuring and appropriately recording contract documentation is crucial when attempting to justify and settle conflict situations. Analysing, meticulously, the piles of project information to sort out conflict/claims is also a thankless task. The sources of project information are many and varied and although the information can be voluminous, it is often inadequate since it contains the wrong information. An advantage derived from keeping track of project information stored electronically is that when conflict arises the information is easily retrieved and disseminated because of the structured systems adopted. Many current paper based information management systems and indeed to some extent IT systems are ill designed to support such conflict/claims events because the information contained within them is either inaccessible or incomplete.

Through the use of sophisticated structured methods such as IT and Electronic Document Management Systems (EDMS), the burden of administration can be reduced and higher quality, relevant documents that can be used and relied upon in a court of law should a dispute arise, will be produced. When dispute or conflict arises the evidence that has to be produced is primarily document based (in most cases paper). Without adequate documentation the dispute will be less likely to be resolved because each of the parties to the dispute will have a difficult time proving the case (without the precise documentary evidence).

Document Management Systems (DMS) have often been viewed as a non-value added component of the construction process: as a result, many construction organisations suffer greatly from the consequences of poor information and administration procedures.

Organisations that support the notion of structured document/information management systems (keeping the most sophisticated documentation records) will have a distinct advantage in any dispute resolution proceeding. Document management systems encourage co-operative working environments and streamlines communication throughout the varying construction organisations. Thus an integrated document management system would support integration and aid the collaborative, integrated team-working culture that the Latham/Egan reports of the 1990s urged. It is highly likely that the use of document management systems would offer new possibilities for the efficient management and conduct of conflict/claims situations that arise within the construction environment as a result of poor or inadequate project information.

4.2.6 Construction documentation in arbitration

In the main, document based evidence is used within the majority of construction disputes that are resolved through arbitration. However, one of the problems facing Arbitrators is poor record keeping and ineffective documentation (Kangari, 1995). When an arbitration case arises it is more likely to be settled if a structured document based information system is in place. Smaller contractors are very vulnerable in this respect since they tend to produce ad-hoc paper based documents. Common problems with arbitration documentation include the following:

- Irrelevant and/or redundant information.
- Excessive quantities of information (volume).
- Poorly structured and indexed information (a structured IT system would automatically prepare structure and index information).
- Lack of information, or worse still, masses of incomplete information.

Kangari, (1995) suggest that many Arbitrators feel aggrieved by the fact that smaller organisations, or those with limited resources, often present their paperwork in an unorganised, unstructured manner. It is clear that those organisations who are able to keep the most comprehensive and structured documentation records have a distinct advantage, compared to the rest, when it comes to settling any arbitration/resolution proceeding and making cogent decisions.

4.2.7 Decision making within the construction industry

Cross-discipline communication between the distinct professions of the construction industry is often problematic and at times non-existent, resulting in poor decision making and reduced performance levels. The demands and expectations of the client are increasing and this has presented a situation where construction organisations are forced to collaborate and share knowledge, skills and information in order to meet the client's demands. This sharing of information is all and well, however, what actually happens to the information once it has been shared and who has the power to make the crucial decisions in order to suit the clients needs often remains somewhat of a paradox.

Within the construction industry, managers (at a range of levels) are commonly called upon to make simple but more often than not, crucial decisions. Managers attach a significant degree of importance to making the correct decisions and often regard this as a key functional aspect of their working day (Fryer, 1998). Many decisions made are snap decisions: made on the spur of the moment and although in the majority of cases these decisions are effective it is a procedure that can be dangerous and in some cases causes delay and disrupt the construction project. Nevertheless a decision has to be made if the situation at that point in time calls for a decision to be made.

Good decision making is often the result of careful analysis and handling of project information. This process has become central to a manager's work: in today's construction industry the ability to pass on information and decisions is extremely vital, although many managers are overcome by the sheer volume of information that is available surrounding the decision. In effect information overload slows down the decision making process because the actual decision making process becomes more difficult to execute since the options and choices are large. The options are limited however if the manager lacks the vital information required to make the correct decision. Organisations are only now beginning to change their working practises as they search for new effective working methods which will aid the decision making process.

Managers are mere mortals and as such they will make errors and mistakes the same as any other human beings might. These mistakes often happen because of external distractions or incomplete information (Wantanakron et al., 1999). Stress is also of concern amongst the industry's managers, and most manager's would agree that the greater number of changes and variations made within a contract the harder the inherent decision making process becomes. This results in more human errors being made that can and will prove costly to all of the construction project's stakeholders.

When errors are made this will inherently affect the way decisions are made. The powers of decision making may well be taken away from those who cannot 'cope' with stressful situations and discharge their responsibilities accordingly. Within the construction industry however, it is necessary to discover what actually causes the errors that may force a manager to make a hasty decision. The following points have been suggested as the main causes of human error, which will ultimately affect the decision-making process.

1. Unfamiliarity – errors occur if a person is not familiar with a situation which happens infrequently. Complex projects and large amounts of design changes will increase confusion.
2. Time availability – not enough time available to detect and correct errors.
3. Understanding – no means available to convey information in a collaborative understandable manner.
4. Information overload – too much information can affect a persons ability to take it all in.
5. New techniques – the need to learn new techniques and change old habits.
6. Quality of information – poor quality, unstructured, inappropriate information affects the ability to make the correct decision. (Wantanakron et al., 1999).

Many of the above errors could be avoided if people within project teams could make group decisions. Group decisions are required when the end decision-maker such as the project manager does not have enough knowledge or information to make the final decision. Group decisions (this may extend beyond the formalised management structure) may be necessary when more ideas and solutions are needed and where more skill and experience is required in order to make the correct decision (Fryer, 1998).

As the projects of the 21st Century become more complex, it is becoming increasingly difficult for one person to make robust and effective decisions. People who are affected by any decisions will want to be involved in the making of that decision. This therefore requires a project team that can collaborate with each other using systems that facilitate this task. A growing number of construction companies are only now implementing IT applications that improve operational efficiency (Molad and Back, 1995) and enable the construction team to make sound decisions based on better more structured information.

As the construction industry's complexity increases, due to the unique design of projects, collaborative IT systems will become an important enabler and allow construction organisation's to perform traditional processes more effectively saving huge amounts of time and resources in the process. The key to effective decision making is that a client and a contractor form a long term partnering agreement in which the parties involved adopt and implement sophisticated IT systems.

4.3 SUMMARY

Over the last decade, the construction industry has seen an emergence and blossoming of interest in applicable IT tools. Through the increased adoption of these tools, construction organisations are beginning to realise the potential they harness. These tools have the power to totally alter current information management processes enabling the sharing of real time information and improving the overall quality of project processes and returning a healthier margin to the bottom line.

This chapter has shown that improved decision making and the better management of project documentation has the potential to reduce conflict and claims situations. A culture where there is openness and willingness to share information is also seen to be conducive to the overall success of the project and ultimately, the industry. The solution is for systems that manage information and also make information more understandable which results in fewer management errors. The construction industry is at last beginning to understand the role IT tools play in the exchange and control of project information. The end result, being a modern construction environment that is collaborative, which in turn reduces conflict due to better communication, quicker transferability of information and an enhanced decision making process.

However, having said all that, we must begin however to comprehend the bigger picture and discover what knowledge is actually being created within construction activity, how it is captured, how it is stored and how it is disseminated through the composite supply chain. An understanding of this and how the use of IT will aid knowledge capture will be discussed within the next chapter.

4.4 QUESTIONS TO PONDER

- The approach to information management is often shaped by the contractual relationships found within the project or operating within the organisation. Is there a single best approach to managing information?
- Information is a people centred resource within the organisation: how do we best extract this information from each individual?
- We all see ourselves as effective communicators, to what extent do we share the same communication base and how does this base manifest itself and be managed within the organisation?

CHAPTER 5

Capturing knowledge within the construction industry

This chapter will consider:

- An introduction to knowledge management.
- Where corporate knowledge is located.
- How we capture knowledge within the construction industry.
- The role of IT in feeding the construction supply chain.
- Knowledge at the construction project level.
- The use of IT as a knowledge capture tool.
- Knowledge capture, storage and dissemination.
- How knowledge aids construction organisational learning.
- Improvement of organisational performance.
- A summary.
- Questions to ponder.

5.1 INTRODUCTION TO KNOWLEDGE MANAGEMENT (KM)

Knowledge Management (KM) is the way that organisations create, capture and utilise knowledge to achieve desired organisational objectives. Within the construction industry it may be argued that we have a limited understanding of what underpins any organisation's ability to harness and exploit knowledge more effectively than its competitors. This is in part due to the fact that knowledge and learning cannot be treated as commodities; they are neither readily identifiable nor measured: they are inextricably linked with the way organisation and the people within it think, work and interact. The drive to increase knowledge management very often means that interaction becomes strained and the individuals concerned see the KM initiative as one more stressor within the work environment.

The key challenges, at the practice level, arise from the need to bring about changes in managerial and organisational behaviours required to constitute learning and knowledge sharing as a productive resource for the organisation. This has implications for the design of strategies, structures, technologies, work processes and also, for the underpinning fabric of social relationships which emerge from, and provide the context for KM's development within any construction organisation.

KM can be viewed as a systematic process of discovering, choosing, arranging, refining and presenting information in such a way that it improves an employee's comprehension (reduction in confusion and stress) relating to a specific area of interest. Nonaka and Takeuchi (1995) classify knowledge in two broad categories: tacit and explicit. Tacit knowledge tends to be personal and hard to communicate while explicit is knowledge that can be transmitted by formal or systematic language.

Within any organisation, KM can be said to have the same degree of importance as labour, plant and materials (Fernie et al., 2003). Some would go further and argue that knowledge is now regarded as the 'primary economic resource' within an organisation (Fong, 2003). Whilst the underlying definitions and approaches to enterprise knowledge will gravitate around the balance of explicit and implicit knowledge, the knowledge still predominantly resides within the brains of the workforce.

KM is linked to numerous processes such as: IT, information systems and e-business, all processes which interact with and influence effective supply chain management. It is therefore argued that, with the potency of KM, the effective management of the knowledge sharing process is tantamount to the successful management of the composite construction supply chain (Ribeiro and Lopes, 2001).

5.2 WHERE CORPORATE KNOWLEDGE RESIDES

Knowledge sharing can be seen as a vehicle to promote the effective integration between members of the supply chain. However we must first begin to understand what form knowledge currently exists in? Xerox (2002) indicate that knowledge is currently held in four main repositories. These four repositories are:

1.   12% of knowledge exists in a shared electronic environment.
2.   22% of knowledge exists in personal devices such as PC's.
3.   24% of knowledge exists in a locked paper environment and:
4.   42% of knowledge goes home at the end of the day.

These figures highlight the underlying fact that most knowledge is held within very specific repositories i.e. the brains of key personnel. This knowledge is incapable of being made available for use by anyone else because it is held in a non-transferable format (would you want a full frontal lobotomy?). It can only be passed on by the individual placing it on some form of transferable medium of by face to face communication. Indeed over 64% of knowledge held in two of the four main repositories displayed above can be attributed to personal/personnel knowledge and of the remaining knowledge, 24% is held on paper which means only 12% of knowledge exists in shared electronic environments such as document management systems or project extranets.

Figure 5.1 highlights a number of issues which can be listed as:

1.   The lack of collaboration between different methods of communication.
2.   The power of the individual.
3.   Little use of industry collaborative platforms such as EDMS/Extranets.
4.   A heavy reliance upon traditional methods of communication (paper).
5.   A lack of tools that transfer paper documentation.

Figure 5.1 further illustrates that the emphasis on managing knowledge needs to move away from managing paper based and personal knowledge to managing knowledge electronically. The establishment of an effective KM strategy enables the organisation to better deal with cultural issues such as resistance to change which can be identified and dealt with in an appropriate manner. As an aid for those individuals and organisations charged with becoming accountable and responsible to their specific area of knowledge, IT provides an enabling platform for across the construction supply chain. Construction organisations that adopt IT as a capture device for knowledge stand to gain or maintain a competitive edge over rivals. The role that IT can play in relation to the capture of knowledge can be of significant advantage to a construction organisation.

As awareness and specific instances of construction IT implementation increases, so greater quantities and varieties of project knowledge will be captured in a meaningful format. This will aid future construction of projects and more importantly increase the knowledge base of construction personnel founded on information created by their predecessors. The use of IT will aid in the coalescence of the distinct professions and professionals who engage in the wider construction industry.

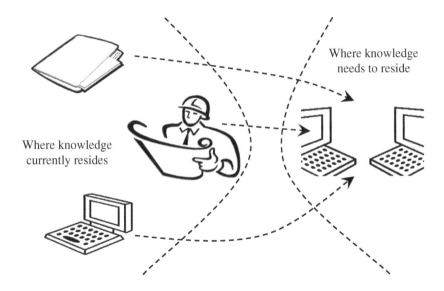

Figure 5.1 Where knowledge currently resides and where it needs to reside

5.3 CAPTURING KNOWLEDGE WITHIN THE INDUSTRY

With the emphasis in today's construction industry, on meeting client's needs, construction organisations tend to look at projects in the short term. Therefore they fail to look beyond the traditional fragmented and non-integrated methods of communication (paper and word of mouth) and ignore KM. The underlying culture, predicated on this traditional approach, ensures embedding of the approach within the culture of the construction industry.

Construction professionals however, all of whom have conflicting priorities and differing objectives, play varying roles and all use their own unique processes to undertake tasks. Crucial to the success of any construction project is the sharing of knowledge between these distinct professions. The distinct construction professionals have become so reliant upon information that has to be supplied by many members of the construction supply chain although the information provided is rarely captured, retained or indexed in a compatible industry wide format. This in itself contributes significantly to fragmentation within the industry and gives rise to many instances of conflict which themselves are very stressful events. The uncertainty and uniqueness of the industry ensures that there is little appeal to the smaller organisations to capture and disseminate knowledge once the construction project is complete since the complex construction supply chain tends to split and go their separate ways. More importantly the knowledge they have created tends to go with them because there is no institution or corpus left where existing knowledge can be accessed unless the project team is part of an alliance or long term partnership such as PFI.

The significant quantities of data currently being created on construction projects emphasises the need for project control mechanisms and has resulted in the construction industry adopting new knowledge capture tools which aid the creation, capture, storage and transfer of knowledge. Organisations that do not systematically secure and capture knowledge created in projects for later use are at risk that the knowledge and experiences gained from within the project will be lost at the end of the project (Disterer, 2002). Kasvi et al. (2003) identify four main areas applicable to KM within a construction project environment:

1. Knowledge creation, for example collection, combination and refinement.
2. Knowledge administration, for example storage and retrieval.
3. Knowledge dissemination internal and external to the project.
4. Knowledge utilisation and productisation, for example integration into products and decisions and application in future projects.

Disterer (2002) argues that the documentation created within projects rarely contains valuable knowledge for succeeding projects, suggesting that project documentation addresses the needs of individuals at the single project level only and can only be a useful commodity on projects of a similar nature (or in the worst case, on this project alone). On larger completed projects, key personnel who often hold the knowledge that will unlock a particular problem cannot be identified or tracked down without considerable effort, through the lack of recorded information.

IT allows for easy identification of key personnel involved in a project and provides a solid foundation for the exploration of new working methods that aid the movement of information and knowledge within the construction supply chain. The issue of isolation and fragmentation can be overcome with the implementation of IT and the dispersed information and knowledge assets can be centrally located for easy accessibility and application by the construction project team. The adoption and application of IT as a KM tool will aid knowledge capture and indeed the IT tools available have proven their worth in the distribution, creation, and management of project information.

Many IT systems are very efficient and effective in the way they capture project knowledge simply because they are implemented and heavily utilised at the most important stage of a project where the bulk of knowledge is created: the construction stage (refer to chapter 9). IT can be used to facilitate the development of innovative ideas during both the construction and maintenance stages of a project. The technology is extremely useful when queried to provide background information for similar projects because the information held provides a good reference point for projects to be planned and executed in an improved manner.

Construction organisations have the ability to share, and more importantly capture, information and knowledge effectively through the use of IT which otherwise could not be achieved if other forms of information storage and knowledge collection systems were used i.e. paper based systems and approaches. As so often is the case with most new technologies, the human element is resistant to change from their existing methods of working to new innovative methods of working, so the implementation process has to involve humans from the lowest level up.

A KM process used in conjunction with sophisticated IT systems will improve the performance of employees and enable them to make better, more informed decisions. Any KM process will also play an important role in re-engineering the culture of the construction organisation towards a culture of collaboration and sharing: this change in culture towards a more open and sharing aspect will in itself act as a knowledge management tool.

Effective KM processes will also generate less stress for organisations and employees trying to do more with fewer resources (Bollinger and Smith, 2001). IT must be put in place to aid the capture of knowledge from existing working processes. More importantly it must be made available to existing and new workers carrying out the same job functions and processes on future projects. IT has become a crucial component for collecting knowledge although how the knowledge is then used, is highly dependent upon the working ways of the organisations within the construction industry. The sharing and dissemination of knowledge throughout the whole of the supply chain is not commonplace within the construction industry more often than not because of a lack of KM protocols.

5.4 KNOWLEDGE STORAGE AND DISSEMINATION

The application of IT as a KM tool can play a vital role in knowledge capture, codification, storage, accessibility and sharing within the organisation. The use of such a system at the construction project level can enable all organisational project information and explicit knowledge to be stored in a central server for members of the organisation and/or project team (supply chain) to access as and when they require it. IT enables knowledge workers to have access to not only important knowledge but the correct knowledge, just-in-time, from other members of the team upon which to make important decisions. It can facilitate faster decision making and also provide a directory of access to key knowledge workers in specialised construction areas. This can significantly affect the value of products and services delivered by the construction organisations to its customers.

IT also allows for easy identification of key personnel involved in the project, thus allowing organisations to contact these key personnel, at a later date, should similar project knowledge and expertise be needed. This stored knowledge (explicit) would include project documentation such as drawings, requests for information, minutes of meetings, specifications, bills of quantity, basic e-mail correspondence and various other pieces of information specific to the project. The use of IT within an organisation and on specific construction projects can enable organisations to share information effectively which would not readily be shared otherwise i.e. if other traditional forms of information storage/repository systems were used e.g. paper based systems. IT can also facilitate the development of innovative ideas both during and after a project since it can be queried seeking background information for similar or new projects. The information held in the system can provide a robust reference point for new projects from which they can be planned and executed, in an improved manner. The role that IT can play to facilitate innovation can be of significant advantage to the construction organisation. The benefits of knowledge sharing using an IT system include:

- Provision of fast, easy access to present and prior construction work that can be re-used for new projects and new ideas.
- Saving of project time by delivering information advice and knowledge 'just-in-time'.
- Availability of essential information at the organisations and employee's fingertips for continued exploitation thus saving valuable time that would otherwise be spent in searching for information.
- Spread and duplication of expertise organisation wide.
- The availability of knowledge for utilisation in the 'future'.

5.5 IT AND ITS ROLE IN FEEDING THE SUPPLY CHAIN

A supply chain is a highly complex network of organisations and suppliers who collaborate more effectively and efficiently to satisfy the needs of the customer (see the earlier discussion on holonic networks in 2.4). These complex networks however depend upon achieving mutually beneficial relationships which embrace dimensions of trust and relationship stability (Mohamed, 2003). The goal of the supply chain is therefore to link all of the supply chain partners and one way to achieve this is the effective use and implementation of IT.

IT is an integral component of any supply chain and is the driving force in any supply chain strategy although the supply chain itself is only as strong as what may be perceived as its weakest link which, no matter what IT system is utilised means the chain can be broken at any time. Many organisations particularly in the manufacturing sector are now members of highly complex supply chains which themselves cannot operate effectively without the aid of sophisticated IT systems (Hammant, 1995). An IT orientated supply chain requires organisations to invest significant amounts of capital although despite the original cost the rewards are even greater.

In the construction sector organisations often appear reticent to apply IT solutions. This reluctance is perhaps based on a degree of uncertainty as to the development and workload prevalent within the industry and from unfulfilled promises made by IT in the past (Motwani et al., 2000).

This uncertainty could be a jaundiced view given the uniqueness of construction projects where teams are formed at the beginning of a project with the goal of undertaking and completing the project within time and on budget. All too often though once the goal has been achieved and with little or no guarantee of future work, the complex supply team that has been formed will be dismantled at (sometimes even before) the end of the project and go their separate ways.

5.6 CONSTRUCTION SUPPLY CHAINS

The effective organisation of a supply chain is essential to the success of the construction project. Construction supply chains are highly fragmented: due in part to the unique entities found operating within the industry and the myriad of professional bodies. These professions, all of whom have conflicting priorities and objectives play varying roles and all use their own unique processes to undertake tasks. Crucial then to the success of any construction project is the sharing of information (knowledge): parties within the construction supply chain however have become totally reliant on information and knowledge supplied from outside sources. This information predominantly being unavailable in a format suitable for all, which further contributes to industry fragmentation.

In view of this underlying fragmentation, it is becoming more important to integrate the various construction disciplines in a construction project with all other members of the project supply chain (Mohammed, 2003). The considerable quantity of project documentation produced on any project is amenable to robust implementation of supply chain management techniques which adopt IT as a tool to control the process/es. As the awareness and acceptance of IT increases, so the members of the construction supply chain will be melded together and coalesce in sharing project knowledge in a meaningful manner.

5.7 CONSTRUCTION PROJECT KNOWLEDGE

The conventional approach to the distribution and sharing of project information has hinged on the use of paper. This of course wastes time and resources and can also be contributable to the rise in construction problems as a direct result of inadequate communication and the poor exchange of information. It is clear that poor information results in poor decision making. Good decision making is often the result of careful analysis and handling of project information and knowledge. Project supply chain teams that use IT are able to share information in an effective manner which ultimately means they can incorporate greater levels of knowledge and power into the project process in order to establish continuous organisational learning and achieve desired (shared) objectives.

Mohamed (2003) argues that the effectiveness of IT within the construction project environment was hampered through the inability of the supply chain members to share information electronically. Advances in IT and in particular web based collaboration systems, has resulted in significantly increased quantities of information being exchanged electronically throughout the construction supply chain resulting in a wealth of knowledge being stored within a central project database. Consideration of web-based systems however has shown that the amount of information contained within these systems is very small when compared to bespoke systems used on one-off projects. Indeed many bespoke systems are very efficient and effective in the way they capture project knowledge: they are utilised at the construction stage where the bulk of knowledge is created. This information can then be accessed and used time and time again to bring about solutions to a variety of problems on many different projects.

The use of IT aids in melding and forging the distinct professions of the construction supply chain into a homogeneous mass. Any supply chain strategy should consider the nature and needs of each organisation and its ability to learn. To learn however, systems (paper and IT based) must be implemented and put in place which capture knowledge from existing working processes and make it available to all existing and future workers. These systems especially IT, are crucial for collecting knowledge although they must be set-up in a way that the knowledge can be easily extracted. How the knowledge is then utilised, is contingent upon the operating methods of each organisation. Organisations that can capture knowledge using IT coupled together with the use of appropriate learning strategies stand to gain significantly from their investment and maintain a distinct competitive over their construction rivals.

5.8 IT AS A KNOWLEDGE MANAGEMENT TOOL

Islands of knowledge amongst construction supply chain partners can become isolated, outdated or lost, thus forcing organisations to reinvent the wheel or do things they have already done. With the implementation of IT this issue of isolation can be overcome and the dispersed organisational information and knowledge assets can be centrally located for ease of access and application by supply chain partners. KM looks at trying to create new knowledge through the capture of existing knowledge and project documentation. The drive being to remobilise existing 'knowledge stock' and deploy organisational knowledge by employing appropriate strategies through the adoption of IT tools. IT is emerging to be a very vital and important KM tool that can address KM initiatives within any organisation.

The precise process by which knowledge can be exploited and measured to meet the needs of the industry will depend upon the KM strategy adopted by the organisation. This strategy will determine how the information is created, captured, stored, shared and transferred throughout the supply chain. As an adjunct, KM addresses issues of identifying deficiencies in the technical standards for IT application and ensures that the facilities are constantly upgraded and compatible with supply chain partners (the use of the Internet alleviates many of these problems). KM also identifies the strengths and weaknesses in not only the organisation but also the individual i.e. skills, competencies and device mechanisms. These weaknesses can be addressed through appropriate internal and external training programmes, the engagement of specialist consultants or by employing appropriately skilled personnel to fill the knowledge gap/s. KM also plays an important and significant role in re-engineering the culture of the organisation making it more conducive for the IT systems implemented to be embraced. Quite often, resistance to change and in particular IT can impact negatively on novel tools and processes designed to improve overall performance.

By establishing an effective KM strategy, issues relating to resistance can be identified and dealt with in advance of the implementation and required change. This will significantly help and aid the integration of the knowledge workers into an environment that is characterised by openness and trust and where IT can facilitate the mutual sharing of knowledge within the organisation. IT can help transform a culture where knowledge hoarding is power into one where knowledge sharing is power which will benefit both the individual and the enterprise. The benefits will also spill over into the wider construction supply chain. IT provides a platform for more transparency within organisations and also across supply chains thus allowing individuals or organisations to be accountable and responsive to their specific area of knowledge contribution.

5.9 ORGANISATIONAL LEARNING: KNOWLEDGE MANAGEMENT

Organisational learning, as a process, can be divided into four stages. These four stages being: (Hoch and Deighton, 1989; Dixon, 1999)

1. Initial acquisition of information; assembling facts, observing and collating data for learning to take place.
2. Interpreting information, producing perspectives, positions, and refining understanding. Questions must be asked at this stage and they are often influenced by the strategy of the organisation.
3. Processing the raw information and reviewing.
4. The final stage is the application of information, engaging in activities and new behaviours. This is when the overall analysis can be converted into action.

A construction organisation's ability to learn from its partners or supply chain is affected by its ability to harness information, and transform/transfer it internally amongst the supply chain as knowledge. Central to the learning process is ensuring that parties within the construction supply chain share similar value systems (culture) and have similar (shared) goals. Learning is predicated on acquiring new insight/knowledge that improves the organisations outcome and performance. Organisational learning is more than the accumulation of knowledge of all the organisations managers and staff, it encompasses also the learning systems implemented (IT) and other processes developed by the organisation to influence the behaviour of others, to establish norms that support a learning environment, and to ensure that the knowledge acquired is passed from member to member throughout the supply chain. The following examples of places where the transfer of tacit knowledge (knowledge in your head) could be openly encouraged illustrate the breadth of potential areas.

- Knowledge interviews, exit interviews, project reviews, report reviews.
- Best practice exchanges.
- Post project reviews.
- Knowledge games and innovation workshops.

While Supply Chain Management can be defined as a value creating activity, many put an inordinate emphasis on tangible measurable outcomes. There exists an apparent tension, often referred to as competitive advantage that implicitly co-aligns partners to learn and acquire new skills, products, technology and knowledge that are unique to the specific relationship. The fundamental challenge is for the supply chain to attempt to learn from others while simultaneously protecting core competencies and retaining its competitive advantage. Learning is consequently influenced by this tension and affected by factors that are relationship specific such as the type of relationship formed and its stage of development.

Organisational learning is also affected by an organisation's absorptive capacity, its willingness and ability to ingest and absorb knowledge from its partners within the supply chain. Even if there is a constant and open flow of communication, information and documentation between the supply chain, some knowledge is tacit and cannot be easily transferred; often impossible to imitate given that it is almost embedded within the people (within the head). Beyond the transfer of tacit knowledge, if internal processes, structures and systems do not encourage information sharing it is highly unlikely that the supply chain will achieve optimal results. In order for learning to take place, several factors may be seen as being crucial, these include:

- A basis of trust and openness.
- A sense of commitment to the ideals of supply chain management.
- Formal communication structures.
- The type of relationship utilised (partnering, alliancing).

The argument is put forward that trust is the key to any collaborative supply chain and depends upon the confidence an organisation holds of others and their commitment, intentions and motivations. Commitment can be demonstrated in a partner's willingness to devote time, energy and/or resources to a relationship and indeed when supply chain members devote resources to ensure longer-term interaction, a higher probability of project success ensues.

Communication is the essential ingredient of commitment and lies at the heart of information transfer. The frequency, depth and content of information communicated have an influence on what is known. Successful supply chain management has been linked to communication frequency and quality and that greater communications lead to better co-operation amongst partners, most managers tend to focus on what is known rather on the how or why it is known (Inkpen, 1998). This explicit knowledge through superficial communications does little to enhance the learning process.

Another factor that affects the learning process is the type of relationship that ties the supply chain partners together. Processes and structures are integrative mechanisms that link the partners at both operational and strategic dimensions through the adoption of appropriate process strategies.

Informal relationships provide an opportunity for deeper knowledge transfer, thereby enhancing the ability to extract and gain from tacit knowledge. If joint activities are performed sequentially rather than in parallel, the knowledge transfer can be better managed and limited to what is needed there and then. It is suggested that IT provides a robust platform from which to conduct such informal activities once the right relationships have been developed and nurtured. This in effect promotes the adoption of greater collaborative arrangements such as learning alliances, partnering arrangements and overall supply chain management.

5.10 KM: AS A CATALYST FOR LEARNING ALLIANCES

Mechanisms that improve inter organisational relationships, commercial relationships and project performances such as partnering and alliancing currently exist within the construction industry. The key to gaining competitive advantage and improving customer satisfaction lies in the ability of construction organisations to form such cooperative 'learning' alliances. Love et al. (2002) proffered that co-operative learning alliances enable construction organisations to better consider business decisions that are influenced by this learning climate, and make best use of the information they encounter. These two concepts impact considerably on transaction costs within organisations. If construction organisations were to utilise a mechanism to best capitalise on these two concepts, then more efficient internal governance structures would result. This can be achieved through cooperative learning where ideas, objectives and mission are shared between alliance partners and throughout the wider supply chain.

Strategic alliances and partnering agreements between two or more firms can enable those partners to learn and acquire from each other, the technologies, skills and knowledge which otherwise may not have been made available to them through traditional working arrangements. Knowledge sharing can act as an enabler for individuals in construction organisations to learn more effectively, especially if operational aspects of the alliance enable continuous improvement activities. The distinct benefits of learning alliances and continuous improvement initiatives complement each other and become more effective when jointly implemented.

According to Crossan and Inkpen (1995), tacit knowledge can be transferred efficiently if the transferor and recipient are linked through common ownership. Common ownership is an implicit feature of strategic learning alliances. It is not surprising to find that many participants who have no power and ownership of the IT system are cautious with the type and amount of information that is supplied towards the system and allowing competitors to obtain a competitive advantage.

Some knowledge (recorded information and documentation) is more transferable than others, and the more tacit the knowledge is the harder it is to acquire. Transfer of this tacit knowledge to a more open, observable explicit knowledge that can be easily learnt is of significant value but is a task which is extremely difficult to undertake. From the construction industry's perspective, the conversion of tacit knowledge to an explicit form of knowledge may create further business opportunities. Organisations must be aware that their members have the potential knowledge for future competition otherwise the potential market knowledge that may currently exist will be lost.

The future challenge for learning organisations in construction is to mobilise knowledge into tools, which can be used to gain a distinct competitive advantage and stay one step ahead of your opposition. The task of turning tacit in your head knowledge into an explicit concept is challenging and involves repeated, time-consuming dialogue and communication amongst project members. It is proffered that the use of IT enhances the creation of such tools.

5.11 IMPROVING PERFORMANCE VIA KNOWLEDGE MANAGEMENT

With the advent of IT and the Internet there is now more room for construction based organisations to compete on both a local and global scale. Indeed the wide spread electronic linking of individuals and organisations (collaboration) has created a new economic environment in which space, time and size are factors which tend to be meaningless. Organisations have the choice to use their knowledge assets appropriately to transform and re-align their operational competencies. The application of any IT within an organisation can effectively facilitate the exploitation of knowledge assets within the project team. In this respect IT can complement any KM initiatives by facilitating knowledge workers' day to day work. IT can facilitate KM to achieve three interrelated goals:

- The achievement of superior external performance, including marketplace and financial returns.
- The achievement of superior internal operating performance through operating efficiencies.
- To enhance the quality of life and culture of the organisation and individual members of the organisation (Fahey et al., 2001).

Application of a KM tool like IT can significantly help construction organisations to innovate new products, services and processes that will reduce the cost and time of delivering these products and services (projects). IT technology enables knowledge workers to have at their finger tips knowledge that has been captured, codified and held in the organisational electronic repository/database. IT through proper KM initiatives can significantly improve the efficiency of the organisation and reduce the time taken to acquire these important knowledge assets. The application of IT is therefore a very important tool for organisations to consider for adoption for the control of their KM strategy and processes.

It is also necessary to reiterate the potential benefits of an improved KM strategy:

- Innovation can thrive where there is a clear KM strategy.
- KM can facilitate the transfer of knowledge between participants, organisations, supply chain, etc.
- Lessons learned can be carried forward on to future projects.
- Firms can respond to clients needs by managing knowledge effectively.
- Organisations retain tacit knowledge which otherwise would have been lost.

- An enhanced knowledge base means that organisations have fewer uncertainties to deal with (Anumba et al., 2005).

The adoption of IT will encourage organisations to upgrade their technical and human capabilities to be compatible internally, with project members and with the wider supply chain. This is a positive aspect for the organisation to be mobilised into a competitive position and better able to respond to internal and external business demands of the modern construction industry. The application of IT can also provide the leverage for the organisations to be able to build and retain their organisational knowledge base, and keep updating their knowledge assets to meet the new and challenging demands of clients and the industry alike. The existing knowledge held in the IT systems will provide the basis in which new business ideas and knowledge can be built on. IT therefore can be seen as an 'explicit knowledge bank' where appropriate knowledge is withdrawn to meet organisational needs.

5.12 SUMMARY

Knowledge management is an eclectic term: in order to extract its maximum potential, concentration on particular facets within its broad constituency is required. IT has a prominent role to play within not only this potential but also within supply chain management. Being part of a supply chain is a huge commitment which involves implementing new processes and strategies to oversee the workings of this chain. This chapter has discussed the applicability of IT to the construction supply chain and has discussed how IT could fundamentally alter the supply chain management process by promoting organisational learning and knowledge management.

The implementation of IT however needs to be reviewed to assess the wider impacts upon knowledge management and organisational learning strategies. For learning to take place it is important that organisations have the ability to capture and assemble information and turn this into re-usable knowledge. IT has been put forward as a solution to this problem in that the systems are more than capable of capturing important knowledge to enable learning to take place.

Furthermore, to improve the quality of project information, management at all levels must become involved, as supply chain management demands a free flow of information between all parties to the project. With increasing focus and fuller awareness now being placed on the customer, the importance of IT and supply chain management within the construction environment is beginning to emerge as a powerful tool for improving the overall quality of projects because the knowledge required to improve the processes has been greatly enhanced by the use of IT.

Finding ways to capture knowledge is one of the drivers behind the call for widespread IT adoption. Linking the knowledge tools available could ultimately result in better and more efficient supply chains where all members have access to accurate knowledge created and distributed by other users within the supply chain. The discussion presented here might guide a more rigorous examination of the linkages between IT, learning and knowledge management. The benefits of utilising the full capabilities of IT within the KM domain may be within reach.

5.13 QUESTIONS TO PONDER

- Given that individuals retain significant quantities of knowledge, how can we best ensure this knowledge is placed in a central repository and made available for sharing?
- If we are to share knowledge, should there be a standard format and who should set the characteristics of this format?
- The implicit assumption in organisational learning is that all individuals want to learn. How can we best manage the situation where individuals have clearly peaked in terms of their organisational learning?

CHAPTER 6

Capturing information at the project level: the reliance on paper

This chapter will consider:

- The reliance upon paper within the construction industry.
- The power and draw of paper.
- The flow of paper information within the construction industry.
- Where paper based information is located.
- Construction – a paper intensive industry.
- Current site information processes.
- Methods of site data capture.
- A summary.
- Questions to ponder.

6.1 INTRODUCTION

The pressures of modern day construction and the increase in client expectations has meant that the construction industry is actively looking to improve quality and performance through increased innovation: one approach to increasing innovation is through project collaboration. Underpinning this concept of construction collaboration is real-time information and communication flow to and from the construction site and the varied contractors. This real-time flow aids the overall construction process and supports data mining and organisational performance management. As briefly mentioned in chapter 4, most construction information is held in a traditional format: on paper. This medium is difficult to access, store and transfer, and creates a bottleneck in the information flow to and from the construction site and its various construction contractors. Current adoption and use of IT systems can in effect increase the rate of paper based information, which is time-consuming and involves a duplication of effort when capturing information because systems are not integrated. The development and implementation of IT systems, in conjunction with dramatic increases in digital based paper technology, will aid the electronic data capture from paper based forms, making the information instantly available across the construction supply chain. These new systems adopted will remove the information flow bottleneck and effectively bring about greater client and contractor satisfaction.

The quantity of data generated increases at near exponential rates at the construction stage (see the case studies in chapter 9). However, we must begin to understand the paper based processes within the industry and how these processes add to the quantity of data created and affect the transfer of communication. Information in a paper based system takes a great deal of resources to gather and analyse which results in an increase in preparation time, an increase in transfer time and a decrease in overall operating efficiency.

6.2 THE RELIANCE ON PAPER

Most construction professionals who have been responsible for a section of or a complete construction project have probably developed some sort of information system (paper or IT) that enables them to carry out their job function effectively. Bowden et al. (2002) indicate that the flow of electronic information comes to an abrupt halt when it reaches the construction site and therefore common construction information systems still remain largely paper based entities where information is unstructured and cannot be used by other construction professionals because it is not easily transferred and, more importantly, becomes lost.

Many construction organisations still have major difficulty in capturing information electronically from the paper source. Existing means of communication are not compatible and the rugged nature of the construction site can result in lost paper documents and damaged electronic equipment (Elvin, 2003). The management of paper based information has therefore become not only a challenge but also a burden upon construction organisations. More often than not, paper based information has to be scanned or indeed re-constructed into a format which can then be readily disseminated throughout the construction supply chain, electronically.

Realistic data entry and the transferring of data is the key to unlocking the power of modern IT tools for all forms of construction management functions (Boehmler, 1998). The adoption of IT would allow for the automatic capture and transfer of paper based information which will be stored and re-constructed into digital, readable and transferable format in an IT database. The ultimate benefit for the construction professional is that they do not have to change their 'old fashioned' ways of working. The approach becomes wholly electronic but still remains a traditional paper based process.

Capturing information electronically can be achieved either by re-entering paper based information into a computer or by using laptop/handheld computers to initially capture the information. The first approach would involve a duplication of effort to capture the information and would be prone to input errors. The second approach using laptop/handheld computers would require rugged hardware, a major change in working practices and a heavy investment in training people to use the system and software in a correct and efficient manner.

The ideal solution to the paper problem leverages new technology in order to automatically capture information direct from paper forms. This ultimately means the construction operative does not have to change the established way of working. They can create construction information and documentation using a familiar method whilst at the same time becoming wholly electronic i.e. in the way the information is transferred and stored. In essence the construction operative becomes fully digital and electronic using traditional paper based processes.

6.3 THE POWER AND APPEAL OF PAPER BASED INFORMATION

The rugged nature of many construction projects renders them as ideally suited to the electronic exchange and interchange of project information using IT tools. However, despite IT being used by a number of construction based organisations to varying degrees of success, the general uptake of IT within the industry lags far behind other industries such as manufacturing (Gidado and Nichols, 2002). The quality of project documentation can be greatly enhanced by using efficient IT systems and although these systems can be seen to be of benefit to site operatives, many of the site processes used directly by the geographically displaced site operatives are still performed via the traditional paper based approach because of the lack of integration with IT systems and a lack of connectivity.

Most project documentation is still prepared and stored as 'hard copy' which is difficult to access and quite often requires vast areas of expensive storage space and more often than not, is the only copy of the document that exists. People feel comfortable with the 'feel' of paper (Bowden, 2002) and the idea of a paperless office is still a long way off.

Figure 5.1 (discussed earlier) demonstrates the industry's players' underlying reliance on paper based information and reinforces the view that too much project documentation still exists in a locked, static, paper environment.

As briefly mentioned in the previous chapter, the emphasis for managing documentation needs to move significantly away from managing paper based documentation to managing the processed data and information electronically. This change in procedure may not be enough because too little information exists in the correct environment 'the shared environment'. In today's construction industry a mixture of methods needs to be utilised for managing information. This mixture of methods includes information produced using IT tools and also information which is created traditionally, 'hand written or drawn' on paper. It is quite often this sketchy information which is of the greatest importance. Information taken down on the spur of the moment can be most useful when dealing with Requests for Information (RFI's) or drawing sketches to send to the architect.

Keeping track of documentation, especially that which has been handwritten, is often a thankless but meticulous task and one which requires a considerable degree of patience. With a paper based system, document control cycles can be dramatically increased as a result of inactive, incorrect and obsolete documents remaining in circulation throughout the supply chain. This leads to inefficiencies and employees following incorrect procedures. Personnel who would like to make changes to documentation often have limited access to the information and ownership of documentation is also difficult to encourage with a paper based system resulting in a lack of accountability and traceability.

The lack of IT tools which can be employed in the analysis and correcting of documentation in paper based systems can make it extremely difficult to find the root causes of incidents: further analysis of paper based information requires greater resources leaving construction personnel less time to spend doing the job they should be doing 'managing and co-ordinating the project'. The difficulties in identifying how and where resources should be directed and allocated can also lead to a rise in costs, an increase in waste, and recurring issues. The solution to the paper based problem therefore is to implement processes and systems which would allow tasks to be 'automated' yet still allow the operatives within the industry to use traditional methods of data capture. This revised process would still allow construction operatives to use pen and paper whilst at the same time turning this handwritten or drawn information into electronic format using the latest IT technology. This adoption of IT technology would undoubtedly speed up the transfer of paper information and has the ability to automatically file all handwritten or drawn information in a secure and shared electronic environment.

6.4 CONSTRUCTION: AN INFORMATION RICH INDUSTRY

The construction industry is an information intensive industry: where huge amounts of documentation are generated and exchanged between the individual (often unique) parties contractually bound to the project. This information intensiveness can also be a barrier to the overall goal of industry 'collaboration'. Indeed the considerable quantities of documentation being created continue to rise; depending on the type of project being undertaken which results in the operatives almost being submerged in information (see the project reviews in chapter 9).

These significant amounts of documentation and the intensiveness of information being created emphasise the need for an appropriate control mechanism and this has resulted in the industry adopting new data capture tools which aid the highly intensive construction processes.

Remarkable advances in information management techniques, and more importantly the use of IT on projects, allows the timely dissemination and transfer of project documentation and communication in a co-ordinated manner. In effect the industry remains as information intensive as ever it was but the information now becomes structured and appropriate for its need. Information integration and collaboration is the key to the achievement of coherent quality management and this can be attained by sharing information electronically during the construction process. However despite the widely acclaimed advantages of using IT such as speed, consistency and accessibility, issues arising relating to time, cost and quality can all be traced back to poor co-ordination and communication which itself is caused by inadequate, insufficient, inaccurate, inappropriate, inconsistent or late information or by a combination of these information factors (Tam, 1999).

Critical to the success of a construction project is the exchange of co-ordinated information although the reality is that construction personnel are drowning in oceans of documents that have been delivered from a range of sources and which are so incompatible that the next stage in the process is to re-create and re-construct the data.

The life cycle approach to modern day contracts, for example PFI, will add to the information intensity problems currently being experienced within the industry. Synchronisation between paper documents and electronic systems has to be achieved if the site personnel at the forefront of the information process are to produce compatible relevant information. Until the intensiveness of paper based information is examined and integrated with modern electronic data capture tools the problem of documentation overload and duplication within the industry will remain.

6.5 QUALITY: THE INFLUENCE OF PAPER BASED SYSTEMS

As discussed earlier, information integration and collaboration are seen as being paramount to the attainment of a coherent quality management approach (total quality as opposed to a functional base which relies simply on the conformity to an approved standard e.g. ISO 9000). This drive for a total quality nirvana can be enhanced through sharing information electronically during all elements of the construction process. However, despite the widely acclaimed advantages of using IT such as speed, consistency and accessibility, issues that do arise relating to time, cost and especially quality, invariably have their roots in poor co-ordination and communication: the genesis of this peril generally being perceived as insufficient, inexact, unacceptable, contradictory or delayed or overdue information, or by a combination of one or more of these.

The main tenets in a total quality approach hinge around everyone involved in the business and the project seeing and agreeing that:

- Quality is what the customer says it is, not what the contractor (or supplier) thinks it should be.
- Conformance to customer requirements is vital.
- Developing systems and procedures to prevent future errors, saves.
- Setting and maintaining performance standards is good.
- Reducing waste is everyone's responsibility.
- Individuals own the right to quality.
- Strong, visible and active support from management is provided.
- Provision of clear, agreed, objectives aids the process.
- Getting it right first time is the goal.
- Effective two-way communications aids in quality attainment.

The implementation of a holistic quality approach can be seen as a continuous process within the organisation and is founded on a 'Six Step Process'. This process provides a framework which should not be viewed as a rigid system but rather, it provides a template which each operating unit can track against, maximising synergistic opportunities wherever possible (Juran and Gryna, 1993; Robertson and Sommerville 1994). The Six Step Process is shown in Figure 6.1, and the constituent steps are:

1. Fact finding i.e. what is the current position of the unit with regard to customer satisfaction; how much waste is there currently; what are staff attitudes? In essence, the identification of need and scope for the quality initiative. Surveys are developed and carried out with clients (these may be internal or external to the organisation/project) to ascertain which products and/or services are of importance to them and within this basket, what they perceive as being the current level of performance towards conforming with their requirements. Staff can also be surveyed with regard to their: job interest, their past and current motivation levels, attitudes to work and the customers, and areas of their work that they and management feel could be improved. A cost of waste calculation should be carried out to ascertain areas where loss is being incurred through money being spent unnecessarily. This provides a clear and consistent picture based on facts, of the external perception and therefore, where major improvements are necessary.

2. Develop improvement plan/s i.e. the considerable quantities of data gathered in stage 1 are presented to a range of participants at a series of quality focus workshops. The workshops encompassing a cross-selection of employees, are aimed at outlining the vision for the future quality state by means of SWOT (strengths, weaknesses, opportunities and threats) and STEP (sociological, technological, economic and political forces) analyses, in order that a coherent and workable quality policy statement can be described and established. This policy statement will then provide the opportunity for everyone to focus on key improvement areas and agree the immediate and near future improvements that have to be made. Highly motivated volunteers are sought out to lead task teams who will be charged with investigating and proposing relevant improvements.

3. Communicate with everyone on the plan i.e. ensure that everyone knows about the commitment to the quality initiative, why they should be committed to it, and what they hope to achieve from its implementation. Step 3 is in essence a detailed communications exercise to ensure that staff who were unable to or failed to attend any of the workshops know what was decided and why. With appropriate IT in place these communications can be rapid, targeted and spread over a wide timeframe.

4. Construct the task teams, with the key improvement areas identified individuals are allocated specific tasks which are clearly articulated to the quality improvement sought. The satisfactory discharge of these tasks and responsibilities demands that the individuals receive appropriate training in a range of quality attainment techniques including: customer satisfaction, communication skills, time-management and, problem solving techniques. Much of this training can now be delivered via online learning resources, so making the learning process more specific, tailored towards the individual's pace and learning style and, more cost effective.

5. Make the improvements i.e. set out statements which include measurable objectives in relation to the identified key improvement areas. With implementation, the improvements are allowed to filter through and subsequent follow-up carried out, to establish the current position in relation to where it was at the outset. Step 5 is where the individual task team members devise improvements, propose these to local steering groups and with approval, implement them, so deepening the sense of ownership of quality. Regular reports, via a range of media, are made on the progress made.

6. Re-commence i.e. after a prescribed period of time has elapsed, the process is begun again with the successes and failures, and important lessons learnt fed back into the overall system. At this stage a one day review of the previous 12 months' activities should be undertaken. New improvement areas may be established and training requirements identified and planned for.

The process of communicating all of these activities and outcomes need not centre on paper. Whilst the ISO 9000 family of standards provides a useful framework for quality attainment, too often it is used to develop a cumbersome paper based regime, which becomes moribund in its own reluctance to move away from paper based audit trails.

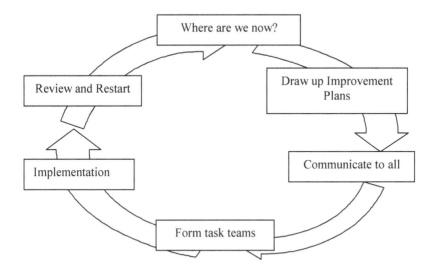

Figure 6.1 The 'Six Step Process' (Adapted from Robertson and Sommerville 1994)

Quality can flourish in environments where paper is not the only medium nor forms the major part of it. What has to be paramount in the minds of those delivering the goods or services are the basic tenets discussed earlier.

6.6 CONSTRUCTION: A PAPER INTENSIVE PROCESS

Current paper based methods of transfer are slowly becoming obsolete because they cannot deliver Just In Time (JIT) information to the construction site at critical stages within the overall process (De la Garza and Howitt 1998). Checklists and standard document templates are very often used to collect and record construction site information and although these current systems are useful for recording data, paper based information is rarely compatible with electronic systems which results in laborious, time consuming procedures, where data and information has to be re-keyed and re-processed into an IT system.

Paper based documentation intrinsically introduces the potential for miss-understandings and miss-interpretations and results in increased risk to the party that created the information. Paper as a form of information has a short shelf life within a construction project and cannot be shared or easily uploaded and updated within an information management system (Boehmler, 1998). Entering paper based information into IT systems is also inherently full of inconsistencies and errors because of the sheer number of stages involved in getting the information into a format which is ultimately admissible in a Court of Law, and also to the correct construction party. It also involves many personnel who do not understand how the information was created or how it is to be manipulated.

Managing project documentation needs to move towards electronic based methods of document control and away from paper based documentation. Within the construction industry however people not only like the feel of paper but prefer to view drawings for example on paper rather than on screen because of the ease of interpretation. A shift towards electronic ways of working will involve some construction process re-engineering as discussed later in this chapter and an examination of which construction processes can be automated.

Before moving on it is worthwhile examining further the problems associated with paper, any potential benefits derived from its use and most importantly, the pitfalls of using paper based documentation. With regards to storage of paper, Meticulus Solutions (2002) indicate that:

- 90% of corporate memory exists on paper.
- Documentation handled within the office is 90% of the time merely shuffled.
- The average document gets copied, a staggering, 19 times.
- 7.5% of all documents become lost and 3% of the remainder get misfiled or misplaced.
- A professional, (which includes all construction specialisms) spends 5% to 15% of their time reading information and up to 50% looking for it.

A slightly different value for each of these is provided by Bowden (2001) who indicated that personnel on a construction site (site and project managers) spend almost 70% of their time dealing with data and this includes generating, managing, sending, collecting and analysing the data. The view of Meticulus Solutions Ltd (2002) that 90% of corporate memory resides on paper is in stark contrast to the view held by Xerox (section 5.2) who indicate that only 24% of knowledge resides on paper. The IT Construction Best Practice Programme (now known as IT Construction Forum) went even further in 2004 arguing that 99% of information is unstructured and held in the form of paper documents.

Given this range of views then, you would expect that the information provided by the site management team to the contractor could be seen as being sufficient, substantial, appropriate and non-conflictual information. On the contrary, the problems of rework occur due to conflicting information and information not received in time. For example, clients very often request a change or a variation in the project and they do not communicate this change to the necessary parties in the time required which results in a rework situation occurring. Newton (1998) indicates that over 65% of contractor rework on construction projects can be attributed to insufficient and inappropriate information with the remaining 35% being split between human error, the weather, workmanship and materials. However, research carried out by the authors has found that over 68% of contractor rework on new housing projects as displayed in Figure 6.2 can be attributed to poor workmanship.

Automating the paper based processes would remove duplication of effort and data input errors: paper based information would become consistent, accurate, appropriate and adequate and would be automatically entered into a back-end database. As stated earlier in this chapter, before we examine current site based paper processes and how these can be re-engineered we require an understanding of the benefits and pitfalls of paper documentation.

The pitfalls of paper as demonstrated in Table 6.1 far outweigh the benefits. What is noticeable from Table 6.1 is that the benefits are 'perceived' which means they are not beneficial to the process but they are perceived to be a benefit by the users of the paper i.e. the humans.

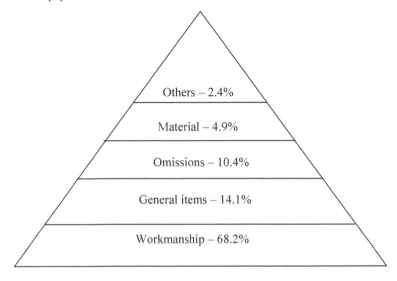

Figure 6.2 The sources of rework on new housing projects

Many paper based systems currently in use within the construction industry are ill designed with little understanding of the actual problem being built into the process. IT can help to alleviate many paper based problems but the IT tools must be designed to match and integrate with current, paper centric, working processes if they are to help and not hinder the information collection process. IT will prove an effective means to facilitate information management: without IT, the process of information management within construction will remain paper based (Stewart and Mohamed, 2003). To determine where this IT technology can be utilised within the construction industry requires an understanding of current site based information processes. Only after these processes have been thoroughly examined will the adoption and integration of paper with IT systems be understood and accommodated.

6.7 CONTEMPORARY SITE PAPER PROCESSES

Current construction site management processes which may be seen as being laboriously slow can be improved by more effective storage and timely communication of information both paper and IT based. The general premise is that this timely communication will tend to be in the electronic sense, whereas in reality many of the site information processes carried out are done so using the traditional method of pen and paper.

Table 6.1 Benefits and pitfalls of paper based documentation (Sommerville and Craig, 2005)

	Pitfalls of using paper based processes	Process Pitfall	Perceived Pitfall
Paper Issues	Poor consistency of documentation	*	*
	Poorly Structured data	*	*
	Unstructured paper processes	*	*
	Increased handling of documents		*
	Poor document quality	*	*
	Poor administration	*	*
	Poor End image/data quality	*	
	Data output	*	
	No data import into database	*	
	Lack of storage device	*	
	No cost savings	*	
Distribution Issues	Loss of paper copies	*	
	Slow speed and transfer of information	*	
	Slower response times	*	*
	Inconsistency with delivery of data	*	
	Poor processing of documentation	*	
	Increased site visits	*	*
	Distribution costs high		*
Legal Issues	No time and identity stamps	*	
	Poor validation of data	*	
	Poor security of process	*	
	Poor safety, reliability of product	*	*
People Issues	Poor communication	*	*
	Poor interpretation of data		*
	Poor decision making		*
	More Physical resources	*	*

	Benefits of using paper based processes	Process Benefit	Perceived Benefit
People Issues	User friendliness		*
	Ease of use		*
	Mobility of paper		*
	Reliability of paper		*
	Familiarity of product		*
	Convenience of paper		*

Within the construction industry there are scores of paper forms and related workflows that employees utilise on a regular basis. The process of completing a paper form, sending it for approval and posting/faxing it back to the client for example is highly inefficient and inconsistent with the electronic way of working especially so when construction site managers, as explained previously, spend almost 70% of their time dealing with and generating project information. Indeed many traditional practises such as 'receipt of drawings in writing' are still observed within the construction industry and the conversion of this process to an electronic way of working has not been widely received. The functions of management personnel on a construction site are wide and varied although the primary functions can be listed as:

- The organisation and co-ordination of contractors and sub-contractors.
- The interpretation of drawings, information and other documentation.
- The allocation, assignment and inspection of work as the project proceeds.
- Administration of all other contract documentation.

To undertake these functions efficiently, the operatives disseminate instructions to the wider construction project team although they are totally reliant upon information supplied by others. Examples of information to be supplied by others are:

- Specifications supplied by the Architect.
- Bills of Quantities supplied by the Quantity Surveyor.
- Health and safety information supplied by the planning supervisor.
- Engineering specifications to be supplied by the engineer.

Many paper based processes within the construction industry (refer to Table 6.2) use pre-formatted worksheets or pro-forma's as a means of data collection. These worksheets however still have to be physically completed and more importantly they have to be re-typed or scanned back into a project database for dissemination.

The complex process of collecting and recording site information can be further hampered if the main contractor or site manager has to disseminate paper based documentation from a number of different organisations (Gyampoh-Vidogah and Nedekugri, 1998). Within these paper based documents that arrive from a range of sources there is no consistency in the approval process. Pre-formatted forms would sit awaiting signing and approval and it is also difficult to track down at what stage the form is at in the approval process. There is also the issue of manually transferring the paper based forms into internal databases where they can be tracked instantaneously.

Efficient site personnel who are able to access and create accurate information will make better use of their project resources. Before this can be achieved however some site process re-engineering will need to take place which should examine which processes can be automated and streamlined resulting in overall improvements to the creation and management of paper based construction site information.

Table 6.2 Site Information processes (Sommerville and Craig, 2004)

Site Information Processes	Mechanism	Operator
Requests for Information	Paper, Worksheets	Architect, Consultant, Contractors
Instructions	Pre-formatted Forms	Architect, Engineer, Contractors
Change Control Proposals	Paper, Worksheets	Architect, Quantity Surveyor, Engineer, Contractors
Records (Site, H&S records)	Pre-formatted Forms	Site Manager, Contractors, Health & Safety Executive
Day Work Forms	Pre-formatted Forms	Site Manager, Contractors, Quantity
Sketches Technical Queries	Notebook, Paper	Surveyors, Architect, Contractors
Impact Analysis Forms	Paper, Worksheets	Quantity Surveyor, Engineer, Consultants
Non Conformance Reports	Pre-formatted Forms	Contractors, Site Managers
Plant Records	Pre-formatted Forms	Contractors, Site Manager, Plant Hire
Attendance Sheets	Pre-formatted Forms	Operatives, All Site Attendees
Snagging Sheets	Pre-formatted Forms,	Contractors, Clerk of Works, Site Manager
Handwritten Notes	Notebook, Paper	All Site Operatives
Site Investigation Reports	Pre-Formatted Forms	Site Manager, Contractors

The objectives of any business process re-engineering exercise is to improve the whole paper and forms processing function, to reduce processing times, to increase paper efficiency and more importantly to improve the service the industry provides to its clients. To highlight the processing time of existing paper based forms we need to examine an existing site process. A Request for Information (RFI) which is a common construction document could for example take a nine step route before approval (see Table 6.3). This nine-step RFI approach is very time-consuming and laborious process and is a process where information could become lost, misplaced or misinterpreted, a fact which will ultimately lead to delays in the overall documentation process.

Notes to Table 6.3:
* Time estimated allows for posting/faxing and transmission of documents of documentation.
** Time estimated allows for creation, transmission and clarification of e-mails.

Table 6.3 A typical RFI process

Step	Stage of process	Estimated Time paper process (days)*	Estimated time electronically **	Stage in process
1	RFI is hand written by the sub-contractor	0.25	0.1	Creation
2	The RFI is then faxed to the general contractor	0.1	0.05	Creation
3	The RFI may be re-written and then faxed to the architect	0.1	0.05	Creation
4	The architect may pass the RFI to a consultant/s for review who in turn	0.1	0.05	Response
5	May pass on the RFI to a sub-consultant	0.2	0.1	Response
6	The responses is are formulated and sent back to the consultant for review	2	1	Response
7	Further review is undertaken and then the RFI is faxed/sent back to the architect	2	1	Response
8	Assuming there is no further clarification needed the architect faxes back the RFI to the general contractor	0.1	0.05	Approval
9	On final approval the contractor passes the RFI to the originator of the RFI	0.1	0.05	Approval
	Total time taken in days	4.95	2.45	

6.8 THE FLOW OF PAPER INFORMATION

A fundamental underpinning to the goal of automation in construction is the flow of real-time relevant data to and from the construction site (Aziz and Tah, 2002). By enhancing the flow of information, it results in the easier monitoring and control of the construction project. The exchange of information electronically can therefore aid the flow of information to and from the construction site and indeed the current paper based pipeline of construction information, which is heavily clogged and currently creates an information deficit, can be transformed into a process which is slick, effective and enhances the organisation's delivery.

On a construction site many different methods and site processes are currently used for creating and managing documents and although the use of EDMS and the adoption of project extranets has increased dramatically over the last few years, paper based information remains the cornerstone of the communication flows. The adoption of these IT systems does not automatically mean the flow of information will improve because many members of the construction supply chain still insist on communicating using paper based documents. Indeed many members of the construction supply are so non IT literate they refuse to transfer to modern methods of working because they feel they add little in the terms of value and efficiency.

The introduction of appropriate IT has the potential to radically alter the current channels and communication lanes within the construction industry. To carry out their job function properly the timing and transfer of information to the construction operatives can quite often be the key to the success of the project. Most use of IT in the construction environment is office based and the construction operatives on-site are often excluded from using IT because the industry's rugged environment does not support the process.

To support the construction operative however and to make construction processes more integrated and streamlined construction organisations are now looking at implementing mobile technology tools on construction sites. These IT tools support the paper based processes although the key objective must be to develop user friendly applications that the construction industry can use to aid management and support the flow of construction information.

6.9 SITE DATA: A SHIFT IN THE METHOD OF CAPTURE

With the demise of geographical boundaries within the construction industry, the mobility of operative's increases as does the distance from their home base office. This spatial; shift requires the industry's operatives to be 'mobile' i.e. migrate from project to project, and also company to company. The industry is inherently peripatetic. To be able to create information on the move requires tools and processes to facilitate this task. Many construction operatives waste valuable time because data has to be captured manually and then transferred back to the office in a physical manner and manually keyed back into an organisations IT system (ICE 2002). Many construction operatives and organisations are now transferring project information using mobile communication tools such as Palm held organisers or Tablet PC's. These forms of mobile communication equipment which are discussed in depth in chapter 7 allow site operatives to examine project drawings and other associated documentation whilst on the move.

Many construction organisations have broken with tradition and are now beginning to adopt these mobile tools as a means of capturing information alongside existing paper based techniques. These mobile tools are revolutionising the information collection process and changing the methods in which site tasks are carried out. However, as we have previously highlighted, industry uptake of IT systems is to say the least slow, so it can be assumed that the uptake of mobile communication tools will not be any faster. Indeed there is little need to use any mobile tool available if there is no IT system for the tool to integrate with otherwise the tool becomes isolated.

The adoption of these mobile tools does bring with it added costs and therefore it is important to manage the appropriate use of the technology in order to maximise the potential benefits the technology will bring (ICE, 2002). The non-adoption of these mobile tools would until recently have meant that the organisations who use the mobile tools would gain certain competitive advantage over the organisations who have not adopted the mobile tools. Now however there is now paper based technology in place that allows paper based processes to be managed electronically and provides a construction operative with the opportunity to turn there handwritten and drawn sketches into digital, electronic format using an old style method, 'pen and paper'. This technology adopts and uses the same technology as other mobile communication tools although put simply it allows paper information to be created, shared and transferred electronically.

6.10 SUMMARY

The underlying communication processes in the construction industry rely heavily on paper. The construction projects themselves repeatedly rely on posted/faxed copies of sketches etc. being transmitted in order that the construction project will continue moving in the planned direction.

Although the use and awareness of IT has increased, many construction organisations still operate and perform tasks in traditional ways. While a number of organisations still favour paper based procurement methods, the method has a plethora of disadvantages. Using this notoriously poor method of communication leaves an organisation open to errors. It is extremely difficult for clients and contractors to obtain up to the minute information and virtually impossible to track it. It is not unusual to observe that certain processes, such as an RFI, can take several weeks to be resolved, when relying upon paper based procurement.

In the future it is hoped that organisations become disenchanted with the use of paper and look elsewhere for a more effective system. However, whilst this book promotes the use of IT we are not extolling the demise of paper, far from it, what we would like to see is the use of paper continuing and promote the use of IT systems that integrate paper electronically (see section 7.11). This concept relies on the integration of a paper based method with fully digital systems. The resulting effect being to create a system which the user sees as being 'traditional' i.e. they still use their pen and paper, and yet is highly sophisticated, from an IT point of view. The sophistication being derived from the ability to capture and transform data first hand.

This chapter has examined the creation of information in a paper context and has also examined current construction site information processes. The benefits and pitfalls of paper based documentation have also been examined and the pitfalls of the paper process far outweigh the benefits. The integration of paper with IT and a discussion of IT tools/systems currently available, and to some extent, in use within construction organisations will be the subject of focus in the next chapter.

6.11 QUESTIONS TO PONDER

- Why do people still prefer to use paper as their main form of information transmission?
- How far are we away from the paperless office given that we have such a range of IT systems and devices?

- The legal profession and the requirement for proof in court cases is moving towards acceptance of electronic documentation. Are the players in our industry willing to accept the move towards an electronic basis or are they dinosaurs?

CHAPTER 7

IT tools: Electronic Document Management Systems (EDMS), extranets and mobile communication devices

This chapter will consider:

- An Introduction into IT tools in the construction industry.
- The use of EDMS within the construction industry, the benefits and features.
- The use of project extranets as collaboration tool, the benefits and features.
- The use and adoption of EDMS and project extranets.
- Mobile communications within the construction industry.
- A comparison of technologies.
- An examination of potential cost savings.
- A summary.
- Questions to ponder.

7.1 INTRODUCTION TO CONSTRUCTION IT TOOLS

The clients of the construction industry, many of whom operate using highly advanced IT systems, are seen as the driving force behind the adoption and implementation of many different IT systems which includes EDMS, extranets and mobile technology. Not content with being the driving force behind the adoption of IT, the clients are also often seen at the forefront of applications take-up.

Construction organisations consume significant time and resources in preparing and distributing project documentation. The use of IT in the construction industry has increased and construction organisations are now beginning to understand how crucial the role of IT is in the exchange of project information. Business use of IT is increasing and equally important is the way in which documents and information is captured and ready available for use for project participants.

The numerous benefits of using IT Systems to control document management have been loosely documented: major savings in time and cost and an increase in overall productivity are only some of the rewards a construction organisation will experience. These benefits however are at best only estimated although a return on initial IT investment can be greatly enhanced if personnel can interact with these pots of information. By analysing the current types of IT technology available we can identify the benefits, the barriers and the key functions of the various IT tools. The drive for value and increasing pressure from clients exemplifies the need for specialised IT tools. It is extremely important to balance the use of such specialised technology in order to balance cost and performance.

This chapter discusses in detail three particular IT tools: EDMS, project extranets and mobile technology (specific emphasis on digital pen and paper) the use of which are now common on the larger construction projects. It is not the objective of this chapter to review all IT tools available. We have placed a specific emphasis on document management and the tools discussed can all be related to document management.

7.2 EDMS: WHAT ARE THEY?

Many terms exist which describe document and information systems such as Integrated Information Management Systems (IIMS), Document Management Systems (DMS), Electronic Records Management Systems (ERMS), Electronic Document Records Management Systems (EDRMS) and Electronic Document Management Systems (EDMS). For the purpose of this book we will use the term EDMS as an all encompassing term. An EDMS has been defined by the National Archives of Australia (2005) as:

An automated system used to support the creation, use and maintenance of electronically created documents for the purposes of improving an organisations workflow. These systems do not necessarily incorporate record keeping functionality and the documents may be of informational rather than evidential value (i.e. the documents may not be records).

The definition implies that EDMS is applicable to organisations on a single basis. EDMS tends to be structured around specific workflow procedures but the systems are capable of providing evidential information and most certainly provide document and record keeping functionality. An EDMS is used to electronically capture, index, store, process, access, view, revise, reproduce, distribute, disseminate information as well as to integrate the information with other relevant project information created in other software packages.

EDMS tends to employ a number of separate technologies each of which can be developed specifically to manage document-based information and workflow procedures within single and multi project environments both in local based systems and web-based format. These technologies include:

1. Document Capture Tools such as scanning, Optical Character Recognition (OCR), Hand Writing Recognition software (HWR) and Electronic Data Interchange (EDI) and automatically create other file formats.
2. Workflow management which support an orderly flow of documents through an organised production process (view the CCP in section 7.7).
3. Groupware systems such as Lotus Notes which provide electronic forms to support flexible collaboration across distinct teams and projects.
4. Document management functions which support organised electronic storage, indexing, version control, archiving, search, retrieval and distribution and dissemination of project documents.

Many construction organisations are now beginning to turn to EDMS in the hope that it will solve numerous communication problems. The adoption of EDMS however does not ultimately mean the communication and collaboration problems experienced will disappear altogether. Indeed many construction organisations still rely on old traditional methods to mange documentation within the project supply chain although major technological developments in the area of EDMS offer the potential to radically change the way construction based organisations create, store and transfer project documentation. There are a number of issues which have to be addressed for EDMS to operate at its optimum level: the main issue, for construction organisations, being that of cost.

Marsh & Flanagan (2000) argue that there are significant barriers which prevent construction based organisations from investing significant amounts in IT. These barriers include uncertainty with the identification and measurement of potential cost savings associated with particular EDMS applications. They argue that the measurement of cost savings is notoriously difficult and the difficulties in identifying the cost benefits quite often prevent EDMS being implemented.

7.3 THE CONTEXT OF EDMS WITHIN THE CONSTRUCTION INDUSTRY

An EDMS tool provides a permanent and highly structured system which can be focused around a single or a number of construction projects. Such a tool will create a central repository of information within a secure environment which allows all project partners to transfer project information electronically fostering collaboration and integration amongst the project team both geographically and temporarily displaced. Many construction organisations have still not taken the first step towards the adoption of EDMS or even IT and it is still possible for firms to exist day to day and compete without using this type of technology. Another reason why the construction industry is not moving towards IT implementation was the lack of good industry specific systems. This has changed however and there are a number of specific industry systems available to implement (this includes project extranets and other IT solutions).

Advances in the development of EDMS and the global nature of the construction industry Worldwide are making the issue of integration all that more critical (Ahmad et al., 1995). Software development or systems development within the construction industry is still highly devoted to fine tuning the systems of individuals. Adopting a wider perspective, which would look beyond the needs of individuals and examines the specific needs of the construction industry, could result in huge competitive advantages for firms as it has in other industries such as manufacturing (O'Brien and Al-Soufi, 1993). EDMS will facilitate the existence of multiple organisations that work in close co-operation and will also allow organisations to improve their operational efficiency; they enable inter-organisational integration of information and achieve reductions in processing critical project information (Molad and Back, 1995).

In construction, the exchange of information is one of the most important functions and the success of this function is highly dependent on the efficiency and effectiveness of the EDMS (Ahmad et al., 1995). A key benefit of EDMS is that it allows organisations instantaneous access to project information. Other main benefits can be outlined as follows with other benefits derived from the use of EDMS shown in Table 7.1.

1. EDMS can reduce the need for bureaucracy and hierarchy.
2. EDMS can be used as a facilitator to build teams and overcome barriers.
3. EDMS will allow flexible communication and eliminate activities or work that no longer adds any value in an integrated environment.
4. EDMS will save participants time and increase their productivity as a result of minimising errors that could normally occur under the paper medium.
5. EDMS increase the quality and speed of work allowing faster simpler access to common data (Ahmad et al., 1995).

EDMS offer the potential to materialise much needed integration within the construction industry. Today, enabled by EDMS it is possible that the construction industry can make effective use of communication, the access of data and common systems to achieve integration within functional levels of all construction organisations. The communication however must be two way, i.e. you put into the system what needs to go in the system and you retrieve from the system the information you need as displayed in Figure 7.1.

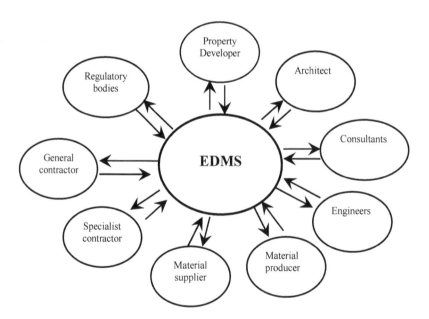

Figure 7.1 Structured EDMS with flow of information internal and external

Table 7.1 The benefits of EDMS from a human, project and organisational perspective

Benefits	Human	Project	Organisational
Reduced printing and distribution costs			*
Effective communications and workflow	*	*	*
Speedier collaboration processes		*	
Improved co-ordination and accessibility		*	
Quick access to all project information	*	*	*
Increased efficiency and productivity		*	*
Comprehensive statistics for benchmarking			*
Secure service		*	*
Information available at end of contract		*	*
Information can be re-used			*
Information security		*	*
The provision of information management	*	*	*
Improved project management	*	*	*
Standardisation of project tools			*
Current snapshot of the project environment	*	*	*
EDMS adds value to information		*	*
Promote standards and best practises			*
Improve workflow processes	*	*	*
Speed and transfer of information		*	
Faster response times	*	*	
Delivery of data	*	*	*
Reduced handling of documents	*	*	
Improved document quality		*	
Consistency of documentation	*	*	*
Improved administration	*	*	*
Enhanced processing of documentation		*	*
EDMS supports quality assurance			*

7.4 EDMS AND SME'S

With a growing emphasis being placed on meeting client needs and improving project performance the delivery of construction projects on-time and within budget is no longer enough (Doherty, 2000). Construction based organisations must therefore look beyond traditional supply chains and create new information and knowledge channels with the wider supply chain members.

EDMS for some time has been embraced by large construction organisations (Stephenson and Turner, 2003) who often work with larger IT budgets. To some extent however the use and implementation of IT can only be justified on larger projects where the cost of such systems can be absorbed into the project value which often excludes SME's from adopting such systems. The aim of any EDMS project is to create an open system which all personnel can use to create and distribute documentation amongst all supply chain partners. EDMS creates not only information but knowledge and can provide others with exactly what they need when they need it (Doherty, 2000).

Demands being made by clients to remain competitive are placing more pressure on the SME's to adopt and implement EDMS solutions that are similar or identical to what is already being used. In effect SME's are being forced to adopt and invest in EDMS solutions. The bigger picture however has not been considered. The best course of action can be to adopt the same solution as the client on a smaller scale although all too often the systems that are put into place fix only immediate problems. The use of such technology would represent a major change in the SME's working ways and without the knowledge of appropriate EDMS solutions SME's will be discouraged and resist the implementation and use of EDMS.

7.5 EDMS AND ITS ROLE IN CONSTRUCTION

EDMS has provided the construction industry with great benefits in the consistency of document generation, accessibility and the exchange of project documentation and the systems have now become a crucial component for managing construction projects. The widespread adoption IT and in particular EDMS has created a culture of openness and trust and one which allows all project participants to have instantaneous access to project documentation although the nature of the industry can also contribute to the lack of IT involvement in many construction organisations (El-Ghandour and Al-Hussein, 2004).

Managing project information with an EDMS removes the onus on the individual to file information; this person in the past had been solely responsible for filing and other associated tasks. The EDM systems should in essence control, store and transfer project information in a systematic manner taking away the need to rely on the individual or organisation to undertake this task. EDMS produce not only vast amounts of information but better and more understandable structured information. In essence the quality of documentation improves irrespective of the quantity.

The accountability and traceability features of modern EDMS tools can also prove to be beneficial when trying to settle complex legal issues that often crop up from time to time within the construction industry because the project consortia would be in possession of the most up-to-date, accurate and admissible project information.

The ownership of project information after the completion of a project is also becoming a major issue as parties' squabble to identify who the owner of such information is. This in turn may affect the ability of legal practises and arbitration courts to settle any legal claims arising from the contract. As organisations begin to change their working processes through the adoption and use of EDMS it is hoped will reduce the amount of risk and uncertainty usually found hiding within the construction environment as a whole.

As the culture of the industry begins to change and embraces technological solutions, the role that EDMS plays will help to reduce conflict and resolve legal disputes. Any legal issues that do arise during and after the contract has finished can be swiftly resolved because the information required to resolve the situation is readily available in a compatible legal electronic format. This seamless electronic collaboration promotes the rapid resolution of on-going project issues (Weippert et al., 2003).

The industry itself has an enormous amount to gain from the application of EDMS and the effective management of EDMS provides real benefits to construction companies of all size.

7.6 THE USE AND ADOPTION OF EDMS IN THE CONSTRUCTION ENVIRONMENT

The introduction and adoption of IT technology such as EDMS has opened the way for construction organisations to reduce the overall cost of creating and distributing project documentation. Many construction based organisations have intelligent EDM Systems in place. The problem is however, that only the organisations themselves have access to the data contained within. Although investment in EDMS has increased, its potential has rarely been realised and many construction based organisations still work with stand-alone systems (ICE, 2002).

In order to improve overall performance and efficiency levels, EDMS usage needs to be significantly increased. EDMS allows all project stakeholders the opportunity to access a central real-time project database. Construction takes place in a very hostile environment. Projects have become larger and more complex. As a result of this, the documentation required to control the overall project process is itself more complex and more importantly of a greater quantity. Being able to share project documentation amongst the project supply chain increases the partnering culture urged by both the Latham and Egan reports of the late 1990s.

Before adopting EDMS as a tool for collaboration it is important to understand the pro's and cons of EDMS. Implementing EDMS can improve decision making and overall effectiveness which under traditional methods of working would not have been available. The main plus points of EDMS adoption can be described as:

1. Fully automated project database with accountability and traceability features.
2. Ease of creation and storage/retrieval of information.
3. Faster response and reaction times to requests for information.
4. Project documentation can be re-used for other projects.
5. The creation of a collaborative environment where access to all information is commonplace (ICE, 2002).
6. Legal issues – many electronic systems and the documentation contained within can now be used to support any claims/legal situations that arise.

The hindrances include:

1. Initial financial outlay – this can be high depending on the size of the project or the size of the organisation.

2. Culture – people like to work the way they feel comfortable. Implementing EDMS changes the way people work. Operatives need to be fully convinced of the EDMS benefits prior to adoption (ICE, 2002).

Effective adoption of EDMS in construction based organisations requires a clear understanding of construction techniques and the overall nature of the business and its operational environment (Then, 1995). The adoption of EDMS can potentially yield cost savings by automating existing processes (Marsh and Flanagan, 2000). Little evaluation of the benefits of EDMS however is carried out and many EDMS or indeed IT systems are put in place just for the sake of 'doing it'. It is difficult to ascertain the potential cost savings of EDMS although the examination of current working processes within an organisation will hopefully identify the areas where greatest improvement can be achieved through the adoption of EDMS.

7.7 KEY FUNCTIONS OF AN EDMS

The management of project information and the need for quicker resolution of conflictual situations are concepts that have driven the development of EDMS. Indeed all types of document management systems and this includes extranets should ensure that critical project information is communicated and shared amongst the system users in a systematic and integrated manner. An integrated EDMS if implemented would also support the claims process, because in the event of a claim a major part of the data gathering is already done (Alkass et al., 1995). An integrated system, which supports the analysis of any claim, is therefore invaluable.

What is therefore required is a system, which supports all types of construction contracts and procurement routes together with a comprehensive change management module which helps to strengthen the decision making process within construction projects. To be effective at the construction stage, the change management module must include the following functions:

1. Change Control Process (CCP) – pre-determined process, which includes the proposal, direction, impact analysis factor (costing), recommendation and final decision.
2. Formal or recognised instruction process (depends on contract type and procurement route being used) that is structured where information flows between specified parties.
3. Site instruction process structured to suit the site management structure.
4. Technical query process which flows as follows: contractor-site manager–architect–site manager–contractor.

The change control proposal is used to create a design change proposal or variation, which is sent to other parties involved in the project for their response and attention. This structured process ending with a final decision /recommendation can be seen in Figure 7.2. This change control process is structured so the client has to make the final decision. This process can also be abandoned after stage three if there is no need to progress the CCP. The Impact Analysis Form (IAF) stage will also highlight any possible savings or extra costs to the project resulting from the design change or variation. This workflow demonstrates how structured EDMS systems support the decision making process in a logical manner.

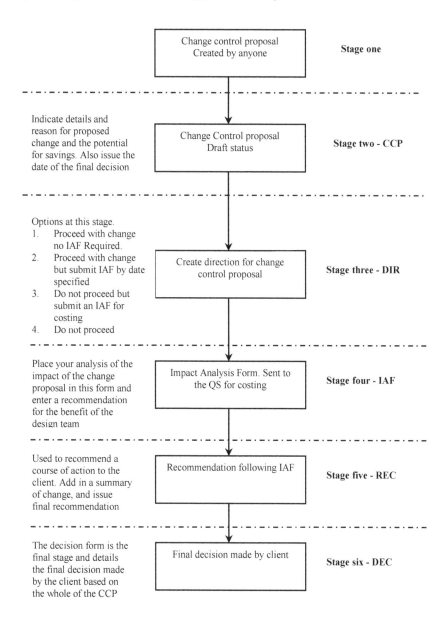

Figure 7.2 Change Control Process within an EDMS (Sommerville and Craig, 2002)

The change management functions are usually found within the change management module of the EDMS. Other modules and parts of the database compliment change management and these are based around general correspondence and drawings. Correspondence tends to be day-to-day e-mails, letters, memos, faxes, discussion documents and contact information. The drawings database is for the handling of all drawing related information. Other key functions of EDMS tend to be more generic and include:

- The ability to link with other databases within the system.
- Access control and login security to ensure access only relevant data.
- The ability to view a wide variety of file formats.
- Workflow ability to control the review of documents.
- A search engine to enable interrogation of the database.
- Multiple notification facility.
- Ability to sort documents into projects etc. using specific codes.
- Request for Information (RFI) functions with complete audit trail.
- Change management facility to handle variations in the design.
- Project address books for all participants.
- Organisational calendar.
- E-mail alerts to documents uploaded.
- History of document access and document creation history.

7.8 THE COST SAVINGS OF EDMS

To stay competitive, construction based organisations must continuously find new ways of working which will improve efficiency and cut costs. Identifying and defining cost savings from benefits derived from the use of IT is difficult but none the less still just as important (Wasek et al., 2000). Much case study data exists within the public domain that defines potential cost savings. However this data is often not quantifiable and at best descriptive.

At a recent IT conference in the UK (ITCBP, 2003), it was noted that a major British supermarket chain were implementing IT collaboration tools by common sense and logic rather than by tangible recorded benefits. They were therefore unable to quantify any financial gain. At the same conference, none of the speakers (industry and suppliers) were able to produce a measured benefit from their adopted IT systems. What is clear is that construction organisations are implementing IT without a clear picture of the potential cost benefits.

What is required is a way of demonstrating potential cost savings which organisations can then compare against the cost of creating documentation using existing working processes. In essence a comparison between the traditional method of creating and distributing documentation and the proposed method after adopting EDMS. As stated however no data exists that quantifies cost benefits derived from the use of EDMS.

A move away from traditional paper based methods may not result in direct measurable cost savings although the elimination of printing, mailing and distribution costs is a major source of cost savings experienced (Wen et al., 1998). In some cases it may be more important to focus upon other benefits such as those which are not measurable. Case studies available within the public domain list many of these non-measurable benefits. These benefits can be described as follows:

1. Client able to monitor document movement.
2. Complete audit trail of all documentation.
3. Postage and telephone call savings (estimated only and not accurate).
4. Man hour savings (estimated only and not accurate).
5. Increased trust and co-operation.
6. Reduction in printing costs (estimated only).
7. Reduction in communication time.
8. Reduces errors and re-work therefore reducing overall costs (estimated only).

Implementing IT without a clear understanding of the potential cost benefits or current working processes is a very high risk process. Many different EDM systems now exist which allow construction companies the chance and opportunity to collaborate and communicate with each other in a manner which is compatible for all. To gain an understanding of the potential cost savings of EDMS, the authors of this book undertook an observation exercise within five construction based organisations who 'DID NOT' currently use EDMS. The five organisations were selected at random from a range of possible participants who had replied positively to an initial scoping questionnaire. The random selection ensured that bias was omitted from the field work and that a representative sample of the underlying population would be utilised.

An in depth questionnaire was used to capture the required data during the observation exercises which took place over a period of approximately 3 hours for each organisation. The questionnaire was prepared with the overall aim of acquiring the working methods of the organisation. The questionnaire itself was split into eleven different sections, each one specifically designed to capture the data for the end calculation of the potential cost benefits. The eleven areas examined and observed were:

1. Paper – amount of paper and cost of paper used per annum.
2. Printing – capital cost of printers and associated products.
3. Computer equipment – functional specification and network capabilities.
4. Existing software applications used.
5. Postage and mail sent on a yearly basis.
6. Filing and storing incoming mail (hard copy).
7. Document access and retention.
8. Document types created.
9. Internet and E-mail access.
10. Telephone calls and faxing.

Specific questions about each section were put to those being observed with general short exercises also being undertaken. The time it took to create E-mails, letters, minutes, memos and file documentation was ascertained. Financial details of the person being observed were also obtained as this detail was vital for the end calculations of the potential cost benefits. After the data was captured it was entered into a spreadsheet and the total figures for the specific exercises were calculated. Some of the figures collected were then used along with the cost of the EDMS to calculate the cost of the same exercises using an EDMS. The end result was the cost savings derived from the use of the EDMS.

Table 7.2 Cost savings derived from EDMS (Sommerville and Craig, 2003)

Observation Company (OB) numbers	OB1	OB2	OB3	OB4	OB5
Generic organisational costs					
Cost of employee working hour =	£9.91	£15.83	£14.31	£13.10	£11.13
Cost per item of postage sent =	£0.36	£0.35	£0.37	£0.74	£0.34
Cost per page of printed paper =	£0.05	£0.06	£0.01	£0.03	£0.02
Cost of a filing cabinet per hour =	£0.49	£0.30	£0.73	£0.48	£0.44
Number of Employees per organisation	17	6	9	64	4
Cost of Creating of a Minutes					
document (4-pg)	2.50	4.08	2.00	7.00	6.00
Creation	£24.79	£64.59	£28.62	£91.70	£66.75
Paper	£0.20	£0.25	£0.05	£0.12	£0.07
Postage	£0.36	£0.35	£0.37	£0.74	£0.34
Filing	£0.33	£0.53	£0.48	£0.44	£0.37
Total cost of traditional method =	**£25.67**	**£65.71**	**£29.51**	**£92.99**	**£67.53**
Cost of creating using an EDMS					
Cost of EDMS per hour =	£0.24	£0.69	£0.45	£0.06	£1.01
Labour Creation =	£24.79	£64.59	£28.62	£91.70	£66.75
Total cost of creation in EDMS =	£25.38	£67.38	£29.51	£92.14	£72.79
Saving between traditional method					
and EDMS =	**£0.29**	**-£1.67**	**£0.00**	**£0.85**	**-£5.26**
Cost of creating a Letter document (1					
page)	0.75	0.20	0.25	0.17	2.0
Creation	£7.44	£3.17	£3.58	£2.18	£22.25
Paper	£0.05	£0.06	£0.01	£0.03	£0.02
Postage	£0.36	£0.35	£0.37	£0.74	£0.34
Filing	£0.33	£0.53	£0.48	£0.44	£0.37
Total cost of traditional method =	**£8.18**	**£4.11**	**£4.43**	**£3.39**	**£22.97**
Cost of creating the using an EDMS					
Cost of EDMS per hour =	£0.24	£0.69	£0.45	£0.06	£1.01
Labour Creation =	£7.44	£3.17	£3.58	£2.18	£22.25
Total cost of creation in EDMS =	£7.62	£3.30	£3.69	£2.19	£24.26
Saving between traditional method					
and EDMS =	**£0.56**	**£0.81**	**£0.74**	**£1.19**	**-£1.29**
Cost of creating a Fax document (1					
page)	0.07	0.05	0.03	0.17	0.08
Creation	£0.67	£0.84	£0.48	£2.18	£0.94
Paper	£0.05	£0.06	£0.01	£0.03	£0.02
Filing	£0.33	£0.53	£0.48	£0.44	£0.37
Total cost of traditional method =	**£1.05**	**£1.43**	**£0.97**	**£2.65**	**£1.32**
Cost of creating using an EDMS					
Cost of EDMS per hour =	£0.24	£0.69	£0.45	£0.06	£1.01
Labour Creation =	£0.67	£0.84	£0.48	£2.18	£0.94
Total cost of creation in EDMS =	£0.69	£0.88	£0.49	£2.19	£1.02
Saving between traditional method					
and EDMS =	**£0.36**	**£0.55**	**£0.48**	**£0.46**	**£0.30**

The results of the observation exercise can be seen in Table 7.2. Examination of Table 7.2 reveals that there is a good range of total employee costs ranging from £9.91 to £15.83, and these reflect the general underlying work-rate in the West of Scotland at the time of the observation exercise. Postage is fixed through national agreements and average organisation costs range from 36p up to 74p. When the range of paper costs were analysed there was significant variance in that the lowest value was 1p per page whilst the highest 500% greater. The cost of filing, storage also had considerable variance with the lowest value being 30p and the highest value being almost two and a half times this sum.

During the observation exercises it was ascertained how long it took to create certain types of documentation. Using the traditional method of creation it can be seen that it costs organisation 4 approximately £92.99 to create, print and distribute a set of minutes. If however organisation 4 adopted the use of an EDMS it would cost approximately £92.14 to create and distribute the minute's document. The savings can be attributed to paper costs and filing. The omission of these elements (which in many organisations is a multiple activity) will give rise to substantial savings on a composite basis. As the organisation population increases so the savings from utilisation of the EDMS increases.

An area which produces substantial is that of creating letters. When the traditional and EDMS methods are compared the best savings amount £1.19 per letter with the lower savings still being an admirable 56p. When the cost of fax documents is considered then again a reasonable saving ranging from 30p to 55p is made on each fax. As a supplementary process area the cost of receipt and distribution of construction drawings shows that adopting EDMS can reduce the cost of producing and distributing drawings by an average of 95% although this information could only be extracted from two organisations as demonstrated in Table 7.3.

Table 7.3 Cost savings from EDMS with emphasis on drawings

	OB3	OB4
Drawings		
Drawings received per year =	1,000	20,000
Drawings Created per year =	50	0
Cost of plotter hardware =	£1,210	N/A
Distribution cost =	£0.60	N/A
Printing cost =	£1.15	
Cost of printing and posting per drawing =	**£1.75**	**£1.00**
Total cost of drawing operations =	£1,840	£20,000
Cost of receiving and distributing using EDMS		
Cost of EDMS per hour =	£0.45	£0.17
Labour Creation =	£0.11	£0.04
Total cost of managing drawings in EDMS =	£0.11	£0.04
Cost saved per drawing =	**£1.64**	**£0.96**

One-off EDMS solutions however have been known to house over one-hundred thousand pieces of documentation and thirty five thousand drawings for a single project valued at £100 million using the PFI route of procurement (see section 9.2.2). The adoption of EDMS is a major decision and the project itself must be able to justify the need for an EDMS system. The type of project such as a PFI or the value of a project will usually determine if an EDMS is required to control and transfer documentation. When current working practises are moved over to an EDMS solution, it is possible to quantify the potential savings. The cost of implementing the EDMS is therefore easily justifiable (Wen et al., 1998).

7.9 SUMMARY TO EDMS

As construction projects become more complex in scope, cost and duration, the establishment of long-term integrated business relationships become cost-effective options for construction organisations and makes partnering and joint venturing a more appealing option for organisations within the industry. EDMS have proven their worth in the creation, distribution and management of project information, and as such provide a solid foundation for the exploration of new working methods that aid the movement of information within the construction supply chain. Current workflow procedures and processes are simplified if the EDMS is matched to existing working methods which, improves the overall operating efficiency of the construction organisation. The use of storage mediums will bring together the construction project supply chain. EDMS supports collaboration and constantly fuels the project with up-to-date documentation which can be utilised by all.

Through the adoption and use of EDMS it is clear that new working methods will aid knowledge capture and bring a wealth of experience to the fingertips of the end user. As the adoption of EDMS becomes more widespread, the cost savings and benefits will become easier to identify. However until the benefits of EDMS are quantifiable construction organisations will continue to work on 'gut feel' and adopt EDMS only for the sake of controlling documentation or satisfying the client.

EDMS will force construction organisations to respond to client needs and the end result could be full-scale adoption of EDMS throughout the construction industry. EDMS and other IT solutions are not the only answer to the construction industry's problems, but will go a long way to promoting integration (Ahmad et al., 1995).

7.10 PROJECT EXTRANETS

Within this book we have tended to focus upon EDMS as a tool for collaboration. It is important however to evaluate and discuss briefly other technologies in the market place that also contribute to project collaboration. Project extranets contribute significantly to project collaboration and have been around for quite a while within the UK construction industry but as a tool for collaboration they did not really take-off until the early part of the decade. The extranets themselves have been transformed from difficult to use and expensive pieces of software into cheap user-friendly collaboration tolls. An extranet is a private network that utilises Internet technology to share project information with partners, sub-contractors and other business associates in order to improve relationship management. There is no unique definition of an extranet as they have been developed from a wide range of systems which were effectively document handling systems. The systems however have been improved over the years to produce powerful functional web based systems that are capable of controlling and distributing documentation on a wide range of projects. Many project extranets offer a common set of functions but many have their own offerings which are unique to that particular system. The common functions tend to be based around:

- Access control – log-in security to view on a status basis only.
- The ability view files without specific software.
- The ability to redline and mark-up.
- Bulk notifications.
- Change management functions.
- Project directories.
- Ability to alert members through the use of other technologies.
- An audit trail and document access history.
- Search engine facility to search the complete system.

The role of extranets in today's business is hard to ignore. Customers can use extranets to access supplier databases, to check on project progress, to issue drawings, and to check information requests. The products are user-friendly, have many functional capabilities and can be justified on both large and small projects.

Unlike paper based systems where documents and drawings are sent to multiple recipients in hard copy format, the extranet stores only one copy of such material which has been 'uploaded' to the system. Once uploaded, team members are notified that the document is available. Controlling documentation in this way ensures that the latest version of the document or drawing is always on the system. In addition to the sharing of project documentation users of extranets are also able to view most formats of CAD drawings without the need for the licence to run particular software. This allows users to 'redline', comment and mark-up drawings and revisions which become important pieces of information within the document chain. This means the likelihood of making mistakes from a project member working from an old paper based document or drawing are in theory reduced or removed entirely.

This method of collaborating working is all done in a closed network, where no-one is allowed access to the system without permission, and everyone is identified. By individually identifying users the extranets are able to automatically track who has seen what, and examine and track any changes made to system documentation. The hierarchical structure of the projects also allows areas of the system to be restricted to only people at certain levels of rank or position in the hierarchy.

The concerns when selecting an extranet is to match the system to the current working requirements of the project and the organisation with the functions and capabilities offered by the various systems (Breetzke and Hawkins, 2003). It is important that an organisation gathers information on available systems, draws up a tight specific specification, matches the vendor's capabilities with the specification and selects and implements the system that is best suited to current working practises.

Besides the electronic business benefits that are commonly experienced, other application benefits of extranets are in information delivery, enhanced communication, productivity, project efficiency and business enhancement to state a few. The top 10 benefits (in order) that companies seek from extranet systems can be described as follows: (Mzenda, 2002).

1. To create a strategic and competitive advantage.
2. To enhance easier access to information within the system.
3. To provide new services.
4. To increase the flexibility of information requests.
5. To improve customer relations.
6. To increase and enhance credibility.
7. To provide better services and products to customers.
8. To increase the volume of information and documentation output.
9. Align well with stated organisational goal.
10. To enable the organisation to respond more quickly to change.

Despite the benefits of extranets there are also many drawbacks such as e-mail overload, copyright issues, cost of drawings to plot to scale, quality and performance of other members infrastructure, incompatible equipment and most importantly the supply chain cannot handle the extranet and prefer hard copy. After examination of the strengths and weaknesses of extranets we must remember that these are IT tools designed to operate in specific areas. With the frantic change of pace of IT technology it is hoped that the use of extranets and other IT systems becomes the norm on construction projects no matter how small they are. The fact that the systems have yet to become the norm could simply reflect the immaturity of the construction sector in relation to other business sectors. Make no mistake, extranets are here to stay but it is important that the IT systems, the supply chain, the contractors, the organisations and most importantly the end users (the people) become that little bit more committed to collaborative working practises that are adopted throughout the construction industry. This can be achieved by adopting extranets as a tool to aid collaborative working.

7.11 MOBILE COMMUNICATIONS IN THE CONSTRUCTION INDUSTRY

Co-ordination and communication frequently breaks down because project information cannot be accessed and transferred at the point of work (Elvin, 2003). Most construction information is still stored as 'hard copy' and because of this is almost unusable at the point of work. Mobile communication tools have made the link between construction site and construction office possible and because of their ability to bring information to the point of work mobile tools may prove to be the missing link in IT (Elvin, 2003). To determine were these tools can be used on-site however requires an understanding of current information processes. Table 7.4 compares the features of digital pen and paper against EDMS systems and standard paper processes.

Table 7.4 Comparisons between paper and electronic processes (Sommerville et al., 2004)

Mechanism	Standard Paper Process	Electronic Process EDMS/Extranet	Digital Paper Process
Storage Cost	High	Low	Low
Level of Scanning Required	High	Low	N/A
Structured Data	No	Yes	Yes
Handwriting Recognition	No	Yes (Limited)	Yes
Reliance upon IT	No	Yes	Yes
Data Transfer	Slow	Instantaneous	Instantaneous
Duplication of Data	Yes	No	No
Reliance Upon Technology	Low	High	High
Hardware/Software Cost	N/A	High	High
Mobility	Yes	Yes (Limited)	Yes
Support Legal Admissibility	Difficult to Prove	Yes	Yes
Distribution Costs	High	Extremely Low	Extremely Low
Consistency of Documentation	Low	High	High
Ease of Use	High	Medium	High

Many construction information processes use pre-formatted paper forms as a means of collecting information although physical activity is still required to complete the form and transfer it back to the office. The use of Personnel Digital Assistants (PDA's) and tablet PC's is increasing and these tools are often used to capture site information. Construction personnel however still love the feel of paper (Bowden, 2002) and therefore the use of these mobile tools as not been widespread. This section of the chapter will therefore examine the use of mobile paper tools that are currently available for use within the construction industry.

Current mobile tools however have been designed with little understanding of user requirements (Anumba et al., 2003). It is therefore of great importance that the mobile tools available are matched to current working methods and improve not hinder the information collection process. In the past, synchronisation between paper documents and an electronic system was not possible. The development of digital pen and paper applications would allow the construction operative to use the traditional paper method for capturing information whilst at the same time synchronising the information with a sophisticated IT system. This section of the chapter will highlight and discuss the use of revolutionary digital pen and paper technology as a mobile tool for collecting site based information.

7.12 DIGITAL PEN AND PAPER IN THE CONSTRUCTION INDUSTRY: THE CONCEPT

Construction information is often locked in a strict paper format which can only be accessed by physically handling the product. There is no opportunity therefore to gain access electronically or remotely to the information held within the paper. To bridge the gap between electronic and paper based systems an automated process is required that allows handwritten paper information to be sent, transferred and more importantly mined electronically. Huge advances in digital technology have provided many industries including construction with the potential to totally revolutionise the way in which paper based information is transferred, stored and disseminated.

The digital pen and paper concept is a concept which has been gaining momentum within the UK construction industry. The technology itself has been developed by Swedish based Anoto and is so unique that it has the potential to totally transform the paper based processes that exist within the construction environment. The concept relies upon other well known technologies such as the Internet and Bluetooth in order that the paper based information is transferred and manipulated.

Manual Process

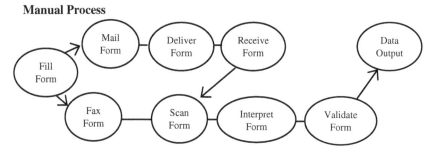

Process Automation Using Mobile Anoto Technology

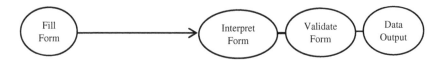

Figure 7.3 Time and resource savings comparison between manual and automated processes
(Sommerville and Craig, 2004)

Two working systems exist and both options will communicate with back-end databases although in different ways. Option one is to use the technology 'locally'. This involves linking the digital pen to a PC. When the digital paper applications are completed using the local options the pen is docked into a cradle attached to the PC and the data is automatically downloaded into the relevant website or database. Much information can be created and transferred this way although the transfer back into the database can only be completed when in the presence of the PC.

Option two allows the user of the technology to operate remotely. The digital pen will record the paper information although instead of connecting it to a PC the pen connects to a Bluetooth mobile phone which then connects to the Internet to transfer the information. The information is then automatically transferred via the Internet and downloaded into an electronic database. Feedback from the mobile phone will indicate if the transfer process has been smooth or whether another connection attempt needs to be made. The process of transferring digital paper based information is dramatically shorter than the traditional transfer process as displayed in Figure 7.3.

Digital pen and paper applications can be linked with a range of electronic databases such as extranets, document management systems, Excel and Access. Further development will bring more functionality and it is possible to turn the handwritten information into digital format and transfer this electronically into different databases as seen in Figure 7.4.

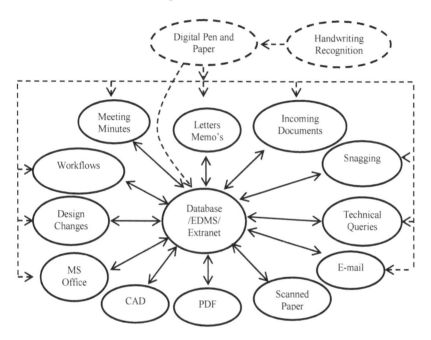

Figure 7.4 Digital paper interactions with documentation and software packages (Sommerville and Craig, 2004)

The additional set-up costs of adopting the technology will depend upon what the size and existing process of the paper application is and the technology to be adopted. The key aspect of the technology is that it is designed to match current working paper processes although the dissemination and data analysis will become totally automatic because the concept is linked into electronic databases. The adoption of digital pen and paper brings obvious benefits although for specific construction applications each process has to be individually evaluated and examined to recognise the potential benefits (Floren, 2002). This must include an analysis of the existing working process in comparison to the new method of working. Only after this has been examined will the decision be taken to adopt the technology within the construction environment.

7.13 DIGITAL PEN AND PAPER AND PDA'S: A TECHNOLOGY AND COST COMPARISON

Mobile IT tools may prove to be the missing link in the success of construction IT because of their ability to link the point of work and the office. The technology to extend IT solutions to the construction site is readily available although construction operatives are not able or unwilling to take full advantage of the potential benefits that the IT tools may bring (Bowden et al., 2002). Many IT tools available are also found to be unusable in rugged site conditions although the technology does provide the possibility for remote access to the information highway from the frontline of construction.

Pocket PC's or PDA's have been in regular use on the construction site for many years. Construction organisations have broken with tradition to adopt the technology although with varying degrees of success because paper is still viewed as the tool to transfer and disseminate information. In order to maximise the benefits of PDA adoption it is important to understand the context in which the technology will be used. It is important to gain an understanding of the type of work being performed, the nature of the work, mobility constraints, the type of data to be processed and more importantly the type of data to be captured before the adoption of such tools (Aziz and Tah, 2002).

PDA's allow mobile operatives to view files such as drawings on the move. Despite this, many drawings are still printed and worked upon in a paper format which eliminates the effectiveness and opportunity for effective feedback from the field of operation (Bowden et al., 2002). Capturing construction site data such as snagging, inspections and piling operations using a PDA is also slow and cumbersome although it does to some extent eliminate the use of paper. However the adoption of digital pen and paper technology allows synchronisation between paper documents and an IT system. Information becomes structured and consistent although the biggest benefit is that the information collected within the remote environment becomes instantly available to the rest of the project team.

The adoption and implementation of any IT tool for use within the construction environment ultimately brings with it added set-up costs. During the feasibility stage of any IT programme it is highly important to be able to compare the cost/benefit of the proposed technological solutions. Floren (2002) produced a report on behalf of the Anoto group which describes the cost/benefit of PDA's in comparison to digital pen and paper. The following assumptions were made:

1. The cost of ownership for the PDA does not include the cost for developing individual customer applications. The digital pen and paper functionality has been estimated on the same basis.
2. User training for the PDA solution is twice (estimated) the time for the digital pen and paper functionality.
3. Support for the PDA is significantly higher compared to the paper based solution although both solutions are based upon the use of Bluetooth technology for operation.

Figure 7.5 highlights a relative cost construct of ownership between a PDA and a digital pen and paper solution. The capital cost, operational cost (solution and technology) and administrative cost for the adoption and ownership of the technologies are all estimated to be significantly lower for the digital pen and paper solution in comparison to the PDA.

The comparison undertaken by Floren (2002) however has been examined with the technology in use in its most basic format. From a general perspective the comparison reveals the attractiveness and benefits of adopting digital pen and paper. However no specific comparison for either technology with regards to the adoption of specific custom built applications has been examined. Therefore until this issue is thoroughly examined the full benefits and cost of ownership of PDA and digital pen and paper solutions and applications will remain at best 'estimated'.

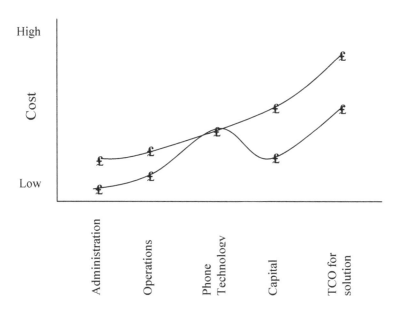

Figure 7.5 Relative cost construct of PDA and digital pen and paper solution

7.14 THE BENEFITS OF DIGITAL PEN AND PAPER

The construction industry is taking a sustained view of how best to adopt and use ICT applications in order to improve construction performance. Historically however the industry has operated on low profit margins which leave little leeway for what may be perceived as high risk investments, e.g. IT solutions which are subject to rapid technological advancement and where a return on investment is not easily measured against an existing process.

If construction organisations could easily measure the business benefits from their IT applications (digital pen and paper is an IT application) then the industry would invest in IT applications. The efficiency, effectiveness and performance of the IT application should if possible be assessed prior to implementation by examining case study material of good and also poor previous implementations. It is also important to assess the performance of the implemented system after implementation and measure the benefits against those specified prior to implementation.

Table 7.5 The benefits of digital pen and paper from technological, process and perceived viewpoint

	Benefits of Adopting Digital Pen and Paper	Technological Benefits	Process Benefits	Perceived Benefits
Paper Issues	Consistency of documentation	*	*	
	Structured data	*	*	
	Streamlined paper processes		*	
	Reduced handling of documents		*	*
	Improved document quality	*	*	
	Improved administration		*	*
	End image/data quality	*		
	Data output	*	*	
	Data import into database	*	*	
	Storage device	*		
	Cost savings			* (note 1)
Distribution issues	Speed and transfer of information	*	*	
	Faster response times		*	
	Delivery of data	*	*	
	Enhanced processing of documentation	*	*	
	Reduced site visits			*
	Distribution costs lower	* (note 2)		* (note 2)
Legal issues	Time and identity stamps	*		
	Validation of data	*		
	Security of process	*	*	
	Safety, reliability of product	*		*
People Issues	Ease of use			* (note 3)
	Enhanced communication		*	
	Interpretation of data	*	*	
	Enhanced decision making		*	*
	Mobility and convenience of the pen	*		*
	Less Physical resources			*
	User friendliness			* (note 4)
	Receiving method	*	*	

Note 1* A perceived benefit as cost savings of paper and other associated material have to be measured and calculated for each individual case. Also depends upon current working process.

Note 2* A perceived benefit as distribution costs have to be measured and calculated for each individual case. Also depends upon current working process.

Note 3* Using the technology is exactly the same as the standard pen and paper process.

Note 4* User friendliness can depend upon the actual application and technology utilised.

The benefits of IT have been well documented throughout this book. It is important however to evaluate and identify the benefits that the adoption of digital pen and paper technology may bring to the organisation which will provide an understanding of the potential impact of the concept upon the organisation. Table 7.5 identifies and lists many benefits that can be attributed to the adoption of digital pen and paper applications. Apart from the obvious technological benefits such as speed and transfer there are also major process benefits to be gained i.e. consistency, data input and processing of paper based information. The new adopted paper process can also help to solve any disputes, claims or legal issues that arise as the system offers a full audit trail which stamps and validates data which makes the creator of such data accountable for their actions.

The main benefits that an organisation will be interested in, is that of cost savings. What is noticeable from Table 7.5 however is that any benefit that relates to cost is a 'perceived' benefit until wider data is available to allow detailed comparisons to be made.

The increase in the accessibility of project data achieved through the adoption of digital pen and paper technology shows significant benefits to all parties involved in project delivery. Professions involved at all stages of project procurement have the opportunity to radically improve their performance and how they interface with the wider supply chain. The adoption of digital pen and paper does not radically alter the way in which current paper based processes are operated and is a solution to the problem of collecting information for dissemination on the move. The solution itself is entirely different when compared to a PDA. Adopting the technology however does not eliminate the use of the PDA solution. Indeed the PDA and digital pen and paper solutions can be coupled together to form the ultimate site operation tool.

7.15 SUMMATION TO MOBILE TECHNOLOGY

There is a compelling case for the use of mobile IT and digital paper applications to cut costs and improve efficiency within the construction industry. Although many organisations have advanced IT systems in place a substantial effort is required to keep them up to date whilst at the same time integrating these systems with existing paper based processes.

Recent trends towards the use of mobile communication tools have increased the opportunities towards improving the flow of construction information. There are many ways in which the construction operative can benefit from the use of mobile tools. With the Construction industry still heavily biased towards the use of paper, the use of paper in an unstructured construction environment will continue as the main mechanism for creating and transferring project information.

The adoption and use of Anoto enabled digital pen and paper technology is a solution to the problem of collecting information on standard paper and its use is becoming increasingly more widespread throughout the construction sector. Used in conjunction with handwriting recognition the technology will eliminate the duplication of effort created by the need to input paper based information manually into software packages. This method of automatic data capture will make construction information instantly available to all construction parties in a format that is compatible to all and has the potential to revolutionise the way in which paper information is collected and transferred.

The construction industry at-large will benefit from the adoption of digital pen and paper technology. The convenience of using wireless communication will ultimately result in improvements to productivity and efficiency. There must however be a compelling case to use the technology and therefore an initial analysis of the paper based process will determine whether the benefits of digital pen and paper technology will improve the conventional paper based process. Although the use of Anoto technology is becoming more widespread, it is often a journey into the unknown and therefore further investigation is recommended which should measure the productivity impacts of Anoto based digital pen and paper technology versus the traditional paper based process. Construction organisations that fail to embrace and implement new mobile technologies are in grave danger of being 'left behind'. However if traditional methods of working are to embrace technology it is important that the technology is embraced around tradition.

7.16 SUMMARY

In this chapter we have examined three different IT tools that are commonly found within use on construction projects. We have analysed the benefits of these tools and were possible the cost savings derived from the use of these tools have been calculated although these have been done so without the use of a standard measurement tool which is something sadly lacking.

The impact of the technologies together with the potential barriers to adoption and successful implementation has also been discussed. The focus within the next chapter switches to the actual implementation of the IT systems. IT has been put forward as a solution to many of the problems the industry suffers from with regards to information management, knowledge management and the problems with paper based information. The next chapter will also dispute that the successful implementation of IT cannot be carried out without firstly matching the implementation model to existing business processes and organisational capability. The implementation of IT will be reviewed in detail to assess the wider impacts upon organisations within the industry. A special emphasis will also be placed upon security and the use of ISO 17799 and other quality standards will also be discussed.

7.17 QUESTIONS TO PONDER

- Given that EDMS is widely available, why then does the industry not adopt it as a standard approach for communicating internally and externally?
- Should the cost savings that are possible through the implementation of EDMS and other IT systems be the sole criteria for deciding weather to implement the system or not?
- The uptake of mobile technology usage is increasing within the industry. Will we see the day where the site manager no longer requires an office?

CHAPTER 8

The implementation of IT within construction organisations

This chapter will consider:

- An Introduction into the implementation of IT in construction.
- A simple guide for preparing for implementation.
- The need to review current working procedures before implementing IT solutions.
- An introduction into the legal issues that affect construction documentation.
- A discussion on the use and adoption of ISO 15489: information and documentation records management.
- A discussion on the use and adoption of ISO 17799: code of practice for information security.
- A summary.
- Questions to ponder.

8.1 INTRODUCTION

IT Systems and in particular Electronic Document Management Systems (EDMS) have for a long time been touted as the answer to many paper related problems primarily because they seem to offer an opportunity to exercise control over the ways in which current paper based documents are handled and distributed (Raynes, 2002). Before the implementation of IT and in order to maximise the benefits of adoption it is important to gain an understanding of current working processes and procedures and match these processes and procedures to that of the IT system. An IT system brings together the IT application, the appropriate processes and most importantly the people who will make it work. To alleviate the shortcomings of IT and its cost to the construction industry, this chapter provides a simple guide to preparing for the implementation of IT within construction organisations (the use of business cases as support for the implementation is discussed in chapter 10). The current chapter will consider the mistakes commonly made through implementation and will also feature a discussion on the legal aspects of document management with regards to ISO quality standards 15489 and 17799.

8.2 PREPARING FOR IT IMPLEMENTATION

The non-physical activities which take place within construction organisations tend to be focused around the creation and distribution of project based information and documentation. When choosing and implementing IT, it is vital to understand the overall purpose of the exercise and examine what it is you are trying to solve. The key stages in preparing for IT implementation can be described as follows:

Stage 1: Setting goals and objectives
The setting of clear goals and objectives for any IT project is an extremely important step. It is important to assess the needs of the organisation and also the needs of the individuals who will be using the implemented system. The problems related to existing functions and processes and potential problems that could be created from the implementation of the new system can be identified by consulting existing staff to obtain their views and opinions. An example of an objective may be that: 'the system should be compatible with existing software and provide co-ordination of activities and processes between the systems'.

Stage 2: Review current document and information management processes
In order to determine how the actual IT system will fit within the business it is important to evaluate current working processes (Gyampoh-Vidogah, 2000). Figure 9.1 below demonstrates a typical paper based document lifecycle process. The evaluation of current working processes will involve either the complete re-design of some processes or only minor tweaking of others.

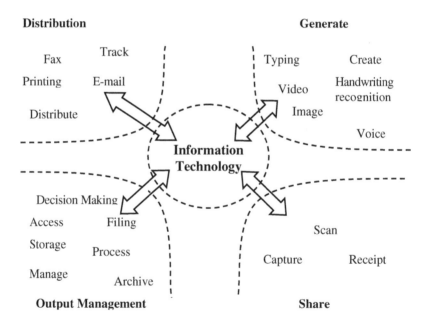

Figure 8.1 The paper based document life cycle process: where IT fits in

This is an extremely important step in the implementation process as there is a danger that the implemented system will be of no core value should things be done in the same way as they always have been. Process mapping is a tool which can be used to identify tasks that add value and support the IT implementation whilst at the same time identifying bottleneck areas that require improvement. The mapping exercise will help to create a strong business case because it will help to remove the non-value adding processes.

Stage 3: Understand the information and documentation exchanged
Construction organisations create and distribute many thousands of documents on a daily basis. When deciding to implement IT many questions require examination when relating to current documentation practises. Questions that can be asked during the analysis stage include:

> How is information currently held and used at present?
> How much information at present exists in electronic format?
> How is the documentation indexed, by job number for instance?
> Who is responsible for document security?
> What happens to incoming paper documents?
> How quickly can filed documents be retrieved?
> How long are documents retained/kept in storage for?
> Are there legal issues surrounding the storage of paper documentation?
> Who has access to the documentation?
> Who is responsible for lost documentation?
> How are documents distributed at the present time?
> How many different document types are there?
> Number of documents created per annum?
> How often are documents up-dated?
> Cost of postage per annum?
> Current document distribution workflow processes (Raynes, 2002).

These are only some of the issues that require examination before implementing IT. Great effort should be made to understand the whole process of document creation. After an understanding of the whole process has been gained, it is then possible to match the existing processes to the working processes of an IT system. An efficient IT system will bring together the processes and the people who use it.

Stage 4: Cost justification of IT
The issue of IT implementation cost and ROI can be a major barrier in the adoption of any IT technology. As discussed in section 7.8, cost savings are incredibly difficult to measure due to the lack of available measurement tools that would assist this task. A decision to implement is often made on a partial cost benefit analysis (Raynes, 2002) although at best this analysis is based upon qualitative benefits and not hard quantitative facts. The situation is also exaggerated because IT providers often make greater claims for their products. Gyampoh-Vidogah (2000) indicates that the costs of EDMS in particular are easy to identify but it is the benefits of the systems that are difficult to measure. The cost driven culture of the industry and the inherently small profit margins realised have a profound effect on the uptake of IT resulting in limited resources available for IT implementation projects.

Stage 5: IT infrastructure required
When a decision to implement IT is undertaken, the choice of system becomes an extremely important factor because this will depend upon the organisations particular needs (Gyampoh-Vidogah, 2000). No single IT tool can meet the needs of any organisations needs and it is common to find a number of systems in use within particular organisations. With regards to the softer IT issues it is important to consider the following with regards to IT infrastructure:

1. What additional hardware/software will the organisation require?
2. Where will the system be hosted?
3. Is the current IT system capable of 'hosting' a new application?
4. Do end-user workstations require upgrading?
5. How much training is needed and over what period?
6. Can the system incorporate existing data as well as new?
7. What back-up systems are in place/needed for disaster recovery?
8. At what times does the system need to be available?
9. What additional IT support is required?
10. What is the security policy for system documentation?
11. Does the system conform to ISO requirements?

After an analysis of the IT infrastructure, other issues need to be factored into the analysis such as user interface requirements, system resources needed and the platform for end user collaboration. These are extremely important factors which could affect not only end user usage but also potential favourable increases in collaboration with members of the construction supply chain.

Stage 6: formulating specific requirements (specification)
After an examination of the IT infrastructure required, enough information will exist that will allow for the formulation of a system specification. In writing a specification it is standard practice to list all of the features and functions that the system should have. Raynes (2002) identifies 4 common areas to focus the specification upon:

Area 1: Performance requirements
This area focuses upon user volume data, number of users, geographical nature of users, volume of data, document types, how often are documents updated, size of documents, location of documents and access to documents.

Area 2: Implementation requirements
This area focuses upon system accessibility, roll-out options, pilot stages, training, data take on (does the system incorporate all existing data or does it start from a new slate).

Area 3: Operational requirements
The focus within this particular area is based around system availability, roles and responsibilities of the IT team within the organisation, back-up requirements, support requirements and system operating procedures.

Area 4: Quality requirements

Quality and security are the main focus within this area with specific emphasis on who is responsible for the security, are there external security issues and what are the system validation requirements?

Stage 7: implementation options

After the specification has been finalised and all options and processes have been examined, the actual suppliers can be short listed and selected through the relevant tendering agreements. How to implement the IT will however depend upon the type of system needed. For example with a single user implementation it would be best to install an off the shelf product. For larger organisations however it may be best to purchase a number of systems and integrate these by developing specific applications. The disadvantage to this approach is the high amount of skills and resources needed to achieve successful integration. Another option is to commission a totally bespoke system although this is only recommended in specialised application areas and should be discouraged (Gyampoh-Vidagoh, 2000). The preferred option for the majority of construction organisations is the selection of a system that meets the majority of their requirements and is a system which can be implemented relatively easily within the organisation. Project extranets and EDMS are systems which fit this purpose and are commonly used within construction organisations although the use and adoption of mobile IT tends to be focused around bespoke systems given its unique ways of working.

The issue of Inter-operability between the many IT systems available is a major challenge facing the construction industry and this may well explain the growth in using web-based applications (systems which require minimal IT development from a user point of view) as an alternative (Stewart et al., 2004).

8.3 MISTAKES AND PITFALLS WITH IT IMPLEMENTATION

Many IT implementations fail and so many go over budget, but why? Best Practice Group PLC (2004) has identified key mistakes in the implementation process that nearly every organisation makes. These mistakes tend to either technical or contractual and are discussed as follows:

1. Requirements writing – if the requirements for the system are written by the organisation implementing the IT then this makes the organisation contractually responsible for mistakes.
   Action: Outline objectives and consult/pay a supplier to understand these objectives and requirements. The IT supplier should things go wrong is then contractually bound to pay for things to be put right.
2. Short-listing IT suppliers – make sure you ask the specialists any technical issues and force them to anticipate and put in place any problem solving measures.
   Action: make sure the organisations business goals form the basis of the tender document and not technical functionality. By specifying objectives the supplier should be able to specify you're your objectives.

3. Insurance – a common problem often ignored is the suppliers insurance cover should things go wrong. If nothing goes wrong during the implementation then there is no need to worry. However should things go wrong and the supplier has no insurance, then the implementing organisation is responsible.

    Action: make sure the supplier has insurance before signing the implementation contract.

4. Clean contracts – before signing any contract agreement make sure all relevant documentation to the implementation project is attached to this document. A clean contract is one that excludes all previous documentation between the organisation and supplier. If you sign a 'clean contract' then all documentation relating to the contract created before hand is rendered useless.

    Action: all documents produced from the ideas, discussions and tendering stages must form part of the contract so the IT supplier understands your needs.

5. Negotiating contract terms – many organisations implementing IT do not really understand the technical and contractual details of the implementation and therefore make very rash decisions. If the organisation does not understand these issues then cut the risk by employing the services of someone who understands the technical and contractual issues.

    Action: always involve a specialised contractual professional during the negotiation stage.

6. Responsibility – many organisation employees become too engrossed within the implementation programme and inadvertently make themselves responsible for mistakes. Take a back seat and do not take responsibility for the IT implementation: this is what you pay the supplier for.

    Action: do not direct the IT supplier ensure the IT supplier directs you.

7. Acceptance windows – after implementation, the IT systems implemented always take time to 'bed-in'. It is essential that during this period any mistakes or functional breakdowns are recorded.

    Action: always build in an acceptance period which will allow you to fully test the implemented system and make any functional changes

8. Specialist vendors – always try and use a 'specialist' supplier. Specialist suppliers will point out what you get and also what you won't. By pointing out what you get/don't get then an organisation will get no financial surprises.

    Action: always use specialist IT suppliers – you then know what to expect.

9. Separate contracts for separate suppliers – During an implementation programme an organisation could potentially have many separate contracts in place. For example, hardware supplier, software supplier, server supplier and infrastructure supplier. Managing these suppliers becomes time consuming and the risk is that the suppliers don't understand your objectives in the same way.

    Action: use only one supplier. Put the onus of separate contracts onto the supplier and this will ensure that your main IT supplier takes the risk.

10. Changing requirements – if you have engaged the services of an expert supplier, then they should be able to define the goals of the project and implement the IT system accordingly. By employing an expert the onus for achieving your business goals and the goals of the system are put onto the expert.
    Action: ensure the expert supplier achieves and understands your business goals. Do not provide them with a shopping list because that is what you will get.

IT implementation failures can be avoided if the ten common mistakes listed are fully understood. It is essential that you outline objectives at the beginning, shortlist suppliers, clarify insurance details, produce documentation for the contract, involve specialised professionals, and let the IT supplier direct you, build in an acceptance period, use specialist suppliers, use only one contract and ensure that the supplier understands your needs. Best Practice Group PLC (2004) indicate that the elimination and avoidance of these common mistakes could mean that the organisation (IT champion) are within the 21% of people whose IT projects have come in on time and more importantly to many organisations on budget.

8.4 BARRIERS TO THE ADOPTION OF IT

This chapter has examined specific areas of IT implemented within the construction industry and focused upon a range of issues emanating from the use of the IT technologies. It is important however to also examine the potential barriers to IT adoption within the industry. Backbolm et al. (2003) found many reasons from industry respondents about why they were not using EDMS on construction projects. The reasons given included: (in no particular order)

1. The respondents didn't know enough about the systems.
2. All participants were viewed as 'local' (no need for complex IT).
3. Current e-mail systems were deemed to be sufficient.
4. Cost of systems was viewed as being too high.
5. A perception of not enough benefit from an EDMS in this project
6. Would like to use but unable to convince others within the project.
7. Simple project, therefore no need for an EDMS.
8. Decision pending.
9. A lack of power in relation to the matter.

The barriers to IT adoption are slightly different within particular segments of the industry. The main industry players, such as main contractors, will use IT in a different way to that of Consultants or SME's. A barrier for a main contractor could be a cultural issue and the actual implementation across the organisation, whereas a barrier for an SME might be training. If you turn these around the barrier for the SME becomes an enabler for the main contractor and the barrier for the main contractor becomes an enabler for the SME. The larger business partners who often work with substantial IT budgets can often demand that SME's that wish to work with them adopt the same systems. In order to meet the demands and expectations of the main contractors and indeed clients, SME's are forced into buying isolated solutions that fix only the immediate problems. SME's need to examine the 'wider picture' and consider the best approach to IT adoption. Other barriers to IT adoption include:

1. Confidence and trust in partners and in the IT systems.
2. Training issues and IT literacy.
3. The change in business processes.
4. Initial financial outlay (especially for SME's).
5. Security issues.
6. Uncertainty of financial returns arising from a lack of reliable, or poor, measurement tools.
7. Traditional communication methods.
8. Lack of management commitment.
9. Technology issues, maintenance costs, IT infrastructure updates.

One answer is to adopt the same systems as your industry partners although on a smaller scale although many SME's because of the slow uptake of IT within the industry have continued operating as they have traditionally done so (Aranda-Mena and Stewart, 2005). The key is to consider and identify what the IT is needed for, select the infrastructure needed and implement the right applications that will fully address the needs of the organisation e.g. an evidence trail.

8.5 LEGAL ADMISSABILITY OF DOCUMENT EVIDENCE

Information generated electronically or stored electronically is subject to scrutiny when required as evidence in a court of law and however information is exchanged there are a number of requirements that any paper based solution and technological solution must fulfil. There are a series of guidelines relating to this type of information; the most important of which for the UK are ISO 15489 (information and documentation records management) and ISO 17799 (code of practice for information security).

As the construction sector begins to embrace IT there is a recognition that as a litigious industry, the importance of legal admissibility of information generated and stored electronically during a construction product is something that must be considered when selecting an appropriate collaboration tool. A typical information system (paper or IT based) must be able to record information in a manner that is understandable and accurate so the information captured can be of use to the professionals that require it.

For legal and professional purposes the retention of documents is a necessity. The storage of paper documents is however non-efficient and businesses continue to look towards the electronic storage of information. A key factor of document retention is being able to prove what actions occurred and when and this can be more easily achieved using a structured IT system instead of having to trail through mountains of paper based construction documents.

Within IT in general, there are some minor problems relating to the storage of electronic documents. There has to be a certainty that if a conflict or claims situation arises that the electronically stored document will have the same weight of evidence in a court of law as the original (RICS, 2003).

Many of the key issues regarding the authentication of electronic information as document evidence include:

1. Originality of the data. Has it or has it not been changed since it was created within the system?
2. Can paper files being imported be reproduced as a true image of the 'original' when transferred from paper into an electronic format through the use of sophisticated scanning devices?
3. The functioning mechanisms of the system?
4. Can the information be retrieved and viewed?
5. Does the electronic system demonstrate the various stages the information went through during its time within the system?

Many IT, and in particular EDMS and extranet, solutions have been developed with a number of features, which support the legal admissibility of information created by or stored within the system including:

- Document Lockdown – documents can be copied and re-used although the original document created within the system can never be altered once it has been authorised.
- Audit Trail – documents should contain a complete history of when they were created and accessed and by whom they were accessed.
- Metadata – documents should hold contextual information about the sender, recipient, filing codes and indexing techniques.
- Authorisation – an authorisation process conveys greater confidence that the data created has been captured and transferred correctly.

In many legal cases the argument is over what a document contains rather than its authenticity (RICS, 2003). Sharp legal minds however will try to discredit the evidence on the basis of its authenticity and it may therefore be necessary to satisfy the court that the information has been stored, copied or transferred in a proper manner or the original may have to be referred back to. Referring back to the original however means a duplication of information but the higher likelihood of conflict within the construction industry forces organisations to keep paper copies and this could be a factor to the slow uptake of technological solutions within the industry.

Construction organisations who adopt IT solutions do so for time and efficiency improvements and the ability to generate savings in paper storage, productivity and duplication. Following the prescribed codes of practice and implementing IT systems that conform to ISO 17799 and improving existing paper based document management systems to ISO 15489 means a construction organisation will be able to demonstrate that the system has complied with regulation and good working practice with the confidence that documents created within the system will stand as evidence in a court of law.

8.6 ISO 15489

Compliance with ISO 15489 ensures that appropriate attention and protection is given to all records and that the information contained within can be retrieved efficiently and effectively. ISO 15489 however, tends to deal with information from a paper perspective. With the construction industry still being so reliant upon the use of paper, we feel it is necessary to discuss the standard in some detail since adhering to this standard could make the implementation of ISO 17799 (should an IT system be implemented) that little bit easier.

The elements outlined within the standard are recommended to ensure that adequate records are created, captured and more importantly managed. Records management governs the practice of records managers and the management of records within an organisation includes:

1. The setting of policies.
2. Establishing guidelines.
3. Designing and implementing appropriate systems for managing records.

The establishment of an efficient records management system also enables organisations to conduct business efficiently, support decision making, facilitates performance, protects its interests, provide evidence of business and most importantly to provide protection and support in litigation. Records management systems are based upon the adoption of specific policies and practises and when implementing such systems it is important that they meet the needs of the organisation and adhere to regulatory guidelines. In summary, a Records Management system that conforms to ISO 15489 should be reliable, compliant, comprehensive, systematic and most importantly, provide integrity should the documents be called upon to support any legal issue that arises.

8.7 ISO 17799

Information and documentation are important business assets and consequently need to be suitably protected given the increasingly connected electronic business environment. As a result of this increased electronic activity, information and documentation is now exposed to a number of threats and vulnerabilities. The security of information can be achieved by implementing a number of controls, policies and procedures. It is important however to ensure that these controls meet the security needs of the organisation.

ISO 17799 is a standard code of practice for information security. However the question remains: why does it matter that an organisations system complies with ISO 17799? As discussed earlier documentation from construction projects has to be relied upon in a court of law although the documents may not be accepted if the legal personnel think the systems that hold the documents are not watertight. Many information systems within the construction industry both paper and electronic have not been designed to be fully secure. However using systems that complies with ISO 17799 will help to ensure that the documentation produced in a court of law is watertight because it has been created, produced, disseminated and recorded in a system that is fully secure.

ISO 17799 creates rigorous audit trails and ensures that documents cannot be tampered with. All access is logged together with activity. Implementing ISO standards can be a long drawn out process and one which often provides little benefit to construction organisations. To date the authors are aware of only 2 IT systems (both project extranets) that comply with ISO 17799.

8.8 SUMMARY

There is a growing awareness of the value that IT can bring to the wealth of parties involved in a construction project. For the industry to benefit however, major cultural changes and an improvement in communication methods must be a real experience so that the people within a project share and transfer information more rapidly and widely, in accordance with their job role. There are a number of operational benefits to be derived from the implementation of a robust IT approach. The benefits however are certainly not assured because they are rarely measured and they depend upon well planned and executed implementation projects.

The requirements to comply with ISO standards relating to electronic document security mean that organisations will have a stronger base from which to ply their case should they be forced to resort to litigation: however, the mere fact that an IT solution is being used should ensure that the risk of litigation is reduced. The key to successful implementation is to understand the needs of the organisation and match these appropriately. Organisations are now turning to electronic storage of documents (records). There is however a number of problems related to electronic storage such as legal admissibility although the storage of documents electronically does not mean it not legally acceptable more likely it is that it will be difficult to prove integrity in a court of law.

8.9 QUESTIONS TO PONDER

- Should all organisations map their business processes and ensure that the individuals within the business understand the operation and integration of these processes?
- What do you consider to be the key factors in the implementation of an IT system?
- Is the lack of IT systems that conform to ISO 17799 a reason for the lack of uptake of IT within the industry?

CHAPTER 9

Implementing IT: reviews of industry projects and applications

This chapter will consider:

- The importance of case reviews.
- The use of EDMS on three construction projects.
- Specific IT implementations within the construction environment.
- The benefits and drawbacks of specific implementations.
- A summary.

9.1 INTRODUCTION

The gathering of case study data has a major role to play in increasing the uptake of IT within the construction industry. The reviews of the cases provide industry with examples of real-life experiences and the benefits gained through implementing IT. How the IT was implemented, and more importantly cost savings which can be reaped, prove of real worth to those contemplating the plunge. This chapter will review case study data from real IT implementations in a range of organisations. The case studies set out to examine the business problem, the existing use of IT, the solution to the business problem, the implementation of the IT, the benefits derived from the implementation and more importantly the lessons learned from the implementation. In addition to this, the chapter begins with a review of three major construction projects and demonstrates the importance of information and document management within the project environment and the scale of the co-ordination problem. The case study data covers a wide range of business processes including EDMS, performance management, customer service management, facilities management, database building, supply chain RFID, digital pen and paper and purchasing procedures.

Whilst other books may concentrate on just the favourable aspects of the implementation process, it was felt important by us that we review the lessons learnt along with the drawbacks of the implementations. It is important to revisit the business process area where the IT was implemented and review its effects, and also to put steps in place to re-engineer areas of the process that still require refinement. Cost savings that are listed within the text are our best estimates: these estimates being driven by the lack of accurate measurement in the case organisations and also by the dearth of models and other tools within the industry and relevant literature, that would allow these savings to be quantified on a more relevant and accurate scale.

The reviews cover three substantial projects in their own rights, and eight applications being developed and applied in specific business areas.

9.2 THREE MAJOR CONSTRUCTION PROJECTS AND THE USE OF ELECTRONIC DOCUMENT MANAGEMENT SYSTEMS (EDMS)

The management and co-ordination of information is crucial to the successful completion of any project. EDMS systems are information management systems where information is passed and shared amongst project members. One particular EDMS has been developed, re-developed and re-defined over time and has been widely adopted as a collaborative productivity enhancement tool for construction projects, and manages the creation, securing, distribution and management of all project data including all correspondence orders and computer generated architectural drawings. The management of project information is a concept that has driven the development of this particular EDMS which itself is a Lotus Notes domino application that can be accessed via Lotus Notes software or an Internet browser. The system ensures that critical information is communicated and shared amongst the system users.

A key feature of this particular EDMS is that it virtually eliminates the duplication of information. The distribution of information within the system, the structure of which can be seen in Figure 9.1 is achieved by sending electronic links to the document, rather than the document itself. By adopting this technique, only one original copy of the document exists placing fewer demands on the human to code and file the information and also less pressure on the project IT system to cope.

Another feature of the system is its unique lockdown feature, which means the original document/s cannot be altered once they have been authorised and distributed. In the event of conflict or confusion within the project the creator of the document then becomes fully accountable for the information contained within the documentation and their actions thereafter. Traceability of the document is instant and the information contained within the system can be located and searched for in a quick systematic manner due to the system's automatic filing and coding structures which are entirely unique to individual projects. This database of information created by project members can then be coupled together with the systems intelligent mailroom feature which allows documents that arrive from outside sources such as E-mail, fax and letter to be routed to the correct recipient or department with the confidence of knowing that those who require to be informed have been notified automatically. The main features and benefits of the document management system can be listed as follows:

- Central repository (database) of project information.
- Faster processing and retrieval of project information.
- Consistency of information between system users.
- Less emphasis placed on the human to file and distribute information.
- Reduced administration and postage costs.
- Project wide tiered access to controlled project information.
- Improved productivity problem solving and decision making.
- 24/7/52 access to the system for all project participants.
- Structured document workflows.
- Unlimited amount of system users. This feature is accessible over the Internet to those who do not wish to sign up to the main central system.

Data analysed from 3 particular projects demonstrates that this particular EDMS provides significant organisational benefits and improves process management. This particular EDMS has been successfully implemented on major construction projects with a combined value of over £6Bn and includes projects such as hospitals, roads, railways and various other projects including PFI projects. The data retrieved from within the three project reviews (all of which used alternative procurement routes) has shown that incorporating and implementing EDMS into current industry practices makes a significant difference to the operation of the construction unit resulting in the benefits described, all of which are valuable benefits to construction companies/clients who adopt the use of such systems.

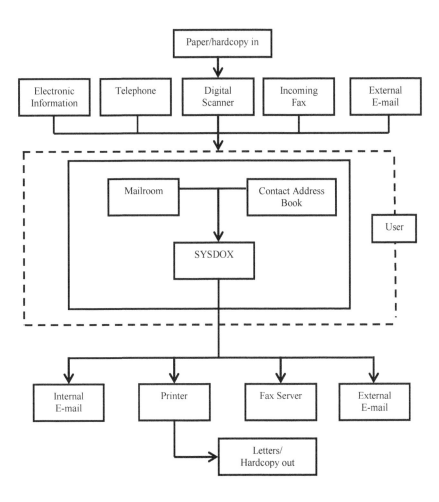

Figure 9.1 The Sysdox Electronic Document Management System (Sommerville and Craig, 2002)

9.2.1 Project review 1: Scottish Executive road building project, overall value £110m

The first project to be examined is a PFI funded motorway project in Scotland which was completed in Early 2000 although the maintenance contract is still ongoing. Broken down into two sections the project involved the construction of a new motorway (M74) (value: £96m) and maintenance of this road for the first five years after completion (value: £14m). The contract was awarded to a collaborative partnering team consisting of four main contractors whose main objective was working together towards a common goal in a spirit of openness and harmony. To assist this harmonious culture, an EDMS was implemented at significant cost allowing all project members access to all project documentation. The scale of the project can be demonstrated by the number of people involved i.e. in excess of 1,000 people on site and 230 system users; it was essential that the system allowed quick access to information from remote locations in a shared systematic manner.

Communication amongst project members could only be achieved using the EDMS. Information that is placed within the system is given a unique reference number, which identifies the creator and distributor of that information. This collaborative information system can be seen in Figure 9.2 below.

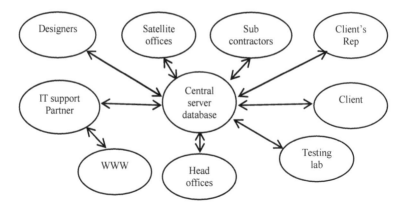

Figure 9.2 Sysdox network infrastructure within the PFI motorway project (Sommerville and Craig, 2002)

The main project partners identified many benefits of using the EDMS collaborative system. These benefits include:

- Accessibility – the system is accessible 24 hours a day providing high-speed dissemination of project information. The collaborative network enhanced the partnering agreement embraced throughout the project.
- Automation – Automated work flow systems shortened the management of processes and tasks. The use of links to documents also reduced the

amount of printing and photocopying required and the storage required for these purposes.

- Quality – The system was identified as a quality 'enhancer' with real time information generated at the click of a button. The system is fully auditable and also makes people accountable for their actions.
- Security – Persons could not use the system without any 'ID'. This ID allowed managers to audit what a person had been working on and what they had created.
- Consistency–due to the databases similar appearance the information was consistent irrespective of the location of the system user.

The distribution of correspondence forms from Table 9.1 demonstrates the amount of information that can be generated on a construction project. The most significant statistic is that over 3000 contacts were listed in the project database, which in real terms could mean there are over 3000 potential users of the EDMS on this particular project. The amount of contacts generated and shared amongst system users also compliments the openness of the partnering culture adopted. Another statistic that stands alone is that there were around 7000 completion certificates issued. This fact is hardly surprising however when one begins to understand the geographical scope of the project (over 90km of existing motorway to maintain and 28km of new build).

Table 9.1 PFI motorway project correspondence statistics (Sommerville et al., 2004)

Document type	Documents (Number)	Document Type	Documents (Number)
Telephone Notes	78	Site instructions	7
Meeting Minutes	1855	Final design drawings	2094
Memos	2691	Method statements	122
Letters	7157	Non-conformance reports	3191
Incoming Documentation	18329	Project certificates	7000
Faxes	3076	Quality procedure docs.	111
Approvals to proceed with final negotiations	20	Requests for inspection of works	2262
Confirm verbal instructions	3	Site photos	408
Engineers instructions	1	Technical queries	1670
Instructions to sub-contractors	1276	Project contacts	3000
Requests for information	11		
Total	**34497**	**Total**	**19865**
Overall Total	**54362**		

Potential cost savings on the new build section alone were estimated at:

- Main site photocopier – only one photocopier required. Estimated savings in the region of £50,000.
- Management savings – if each user saved 1 hour per week then the potential saving in management time over the duration of the project is estimated at over £300,000.
- Programme savings – rapid distribution of information resulted in fewer delays to the contract as a result of awaiting information. The estimated saving was over £200,000.

The findings from this particular project review clearly indicate that major contractors and project partners have come to rely on the EDMS as their main source for project information. The collaborative partnering culture was essential to the success of the project and produced vast savings (although only estimated), which further demonstrated the benefits of using EDMS. The information within this particular document management system was structured in a systematic manner and in a way which ensured all project participants had instant access to all project data. The system also transferred information to remote users which helped to speed up the decision making and problem solving process which eventually lead to the project being finished ahead of schedule. The total overall savings which were in the region of £500,000 was more than sufficient to cover the initial cost of purchasing, developing and setting up the system.

The main benefits accrued to the contractor, who had instant access to all project information because all incoming correspondence and correspondence created by users was easily managed and automatically filled taking away the need for every user to manage project documentation. The sheer size and volume of the documents created also demonstrates just how big a task it is to manage project correspondence correctly. This has resulted in the free flow of information throughout the project team, which has placed the emphasis on all project participants to effectively manage project data in a way that is systematic for all.

This project review has examined the use and adoption of an EDMS within the construction industry and found that if the construction industry is to continue improving it must adopt a collaborative culture where all project participants have access to all project information. The sharing of project information is essential to any construction project and the EDMS adopted within this particular project demonstrates that an EDMS is more than capable of supporting collaborative open networks within the construction industry. The case review data has also shown that standardised document management systems are the way forward and the evaluation of this particular EDMS as a collaborative management system demonstrates the enhanced effectiveness and efficiency of project delivery.

9.2.2 Project Review 2: PFI Hospital, location Scotland, overall value: £100m

This particular project, completed 2001, in central Scotland, replaces an existing hospital building and was valued at approximately £100m. The project was delivered under a Design and Build Contract regime. Under this regime, the principal contractor is responsible for the design of the hospital and its construction. The traditional role of the Architect was subsumed within the construction organisation. The project delivery entailed some fourteen different organisations linking to supply the finished article-the hospital. The principal contractor saw the benefit of installing an EDMS that would allow information created by the project participants to be used by others both during and after project completion.

The system was enhanced to allow the creation and distribution of project information, which covered all correspondence and most importantly drawings. Also within the system a change management module existed and each document created within this module had built in security features and workflow processes. This made the document unchangeable once it had been placed within the system and forced the receivers of such information to make better, quicker and more effective decisions.

Table 9.2 Documentation statistics from the hospital project (Sommerville and Craig, 2002)

Correspondence form	Documents created	Attachments	Documents per month
Builders Work Request	90	100	3.0
Change Control Proposal	1500	500	50.0
Client Observation Formed	200	250	6.7
Company Record	450	0	15.0
Contact Record	1300	0	43.3
D&B Contractors Instruction	1500	350	50.0
Decision's Made	1500	3	50.0
Directive	1500	0	50.0
Drawing Record	35000	0	1166.7
Fax	3300	1650	110.0
Impact Analysis Form	4200	900	140.0
Incoming Documents	8300	8100	276.7
Letter	2000	600	66.7
Memo	8300	6400	276.7
Minutes	2000	1500	66.7
Site Direction	150	10	5.0
Site Instruction	10	10	0.3
Sign off Notification	550	1115	18.3
Telephone Call Notes	100	10	3.3
Technical Query	4000	2550	133.3
Transmittal	16000	200	533.3
Totals	**91950**	**24248**	**3065**

The Design & Build procurement route chosen explains why there is so many drawing records and change control proposals. This type of procurement route allows for the need to make changes and variations to the contract requirements as work continues to proceed. It is also clear from Table 9.2 above that there are approximately three impact analysis forms for every change control proposal form (for a discussion into these forms refer to section 7.5). Therefore for each change control proposal an impact analysis form regarding this proposed change is sent to three other parties within the contract, for example the quantity surveyor or construction manager. Another interesting statistic is that the amount of change control proposals reflected the amount of actual contractor's instructions given. This demonstrated that the contractor only issued instructions and made decisions regarding changes. Furthermore the amount of technical queries to the principal contractor from other contractors could also explain the amount of change control proposals. The main point is that the whole process was structured (refer to Figure 7.7) and the final decision regarding the change control proposal was not made until all the different possibilities and solutions had been examined. The system also allows operatives the opportunity to track documentation response times. In effect performance managing the supply chain.

These findings clearly indicate that Design & Build projects produce vast amounts of project information and that the contractors relied upon the EDMS as their main source for project information. This standardised information has also improved decision making and the better management of project documentation reduced the conflict and claims culture normally associated with the industry. The EDMS described in this project review has demonstrated how EDMS tools can be used to support decision making and create harmony within the Construction Industry. This project review has also examined the documentation contained within the EDMS used and found there was a staggering 91,000 documents created during a 30 month period. More specifically however the change control process was examined and this was found to be a structured process where decisions are made in an orderly fashion ending with a recommendation or final decision.

9.2.3 Project review 3: refurbishment project

To highlight the further use of an EDMS on a construction project, data received from a completed £20m refurbishment project within the heart of the financial district of Edinburgh, Scotland has been analysed. Document communication on this particular project (which was built using the construction management procurement route) could only be achieved by use of the EDMS, a system which was driven by the client but equally well adopted and supported by the main contractor. Information was uploaded and distributed via the EDMS and features within the system such as audit trail, lockdown, document access history and accountability allowed for the control of documentation in an efficient effective manner.

The EDMS included a correspondence tracker, change management module, configuration module (project set-up) and a drawings module. Actual amounts of and types of information created from the project highlight the workflows adopted within the system and to some extent the amount of integration amongst project partners. Table 9.3 indicates the amount of project information created on the refurbishment project using the 3 main system modules (set-up module not included).

The distribution of information from Table 9.3 demonstrates the amount of information that can be generated on this particular type of construction project. A statistic that stands alone is that there were over 7200 requests for architect's instructions which, demonstrates the type of project that has been worked upon.

Fourteen main design team (project team) members were responsible for the creation of over 82% of general correspondence and received nearly 80% of total correspondence. The most information intensive user was the main contractor who created over and received over 45% of total correspondence documentation. (The figures do not include change management and drawings).

The aim of the EDMS was to create a system which project partners could use to create, transfer and distribute documentation amongst the many project partners or in simple terms 'collaborate with each other'. The EDMS on this particular project was an integral part of the overall supply chain and was the main driving force behind the successful collaborative culture that was created. This collaborative harmonious culture created was essential to the overall success of the project and also indicates that construction organisations that in the past were reluctant to collaborate with other distinct identities are indeed willing to share information and become more integrated and open with each other.

Table 9.3 Documentation statistics on the refurbishment project (Sommerville et al., 2004)

Database Types Used					
Correspondence Tracker		**Change Management**		**Drawings**	
Document Types	**Totals**	**Document Types**	**Totals**		
Archive Documents	1469	Request Architects Instruction	7218		
Fax Documents	2678	Architects Instruction	7273		
File	4	Construction Managers Instruction	7297		
Incoming Documents	5030	Change Control Proposal	468		
Letters	1932	CCP Direction	482		
Memorandums (e-mails)	7829	CCP Impact Analysis Factor	1294		
Minutes of Meetings	665	CCP Recommendation	471		
Telephone Notes	16	CCP Decision	417		
		Technical Query	767	16500	
Total	**19623**	**Total**	**25687**	**16500**	

Project Total = 61810

The findings from this particular project review clearly indicate the sheer volume of information that can be created on a particular type of project. The client, contractor and project team have come to rely upon the EDMS as their main source for accessing and distributing project information. Through the use of the EDMS compatible and structured information has been created and produced that has been transferred, viewed and more importantly disseminated across the distinct project team. Importantly, the use of the EDMS has demonstrated that it is possible for construction organisations to operate in a collaborative manner within refurbishment projects.

9.2.4 Project reviews summary

Vast amounts of project documentation have been created within the three project reviews. However as Figure 9.3 demonstrates, the types of documentation created within different types of project using different contract and procurement routes differ greatly. The project reviews have shown that on projects were design has been finalised then the amount of drawing information created during construction is extremely low when compared to a design and build or refurbishment project. Alternatively, the amount of change management information created during the hospital project reflects the type of procurement route undertaken and this would also account for the extreme amounts of drawing information.

If EDMS tools are to become the norm within the Construction Industry it is important that EDMS tools can be adapted to different contractual situations to ensure information is transferred quickly and efficiently with the guarantee that the information can then be used to make the correct decision (i.e. Design & Build, Traditional, Construction Management). Better decision making is only one aspect of an integrated construction process and implementing systems such as this particular EDMS has reduced conflict and improved the overall effectiveness of the construction projects due to the consistency of decisions made.

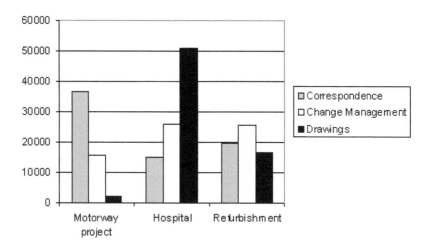

Figure 9.3 Types of documentation within different project types

The identified need for collaborative working is being reflected in the growing usage of EDM systems. A culture where there is openness and willingness to share project information is seen to be conducive to the overall success of the industry. With increasing focus and fuller awareness now being placed on customer quality, the importance of document management systems are now beginning to emerge as a powerful tool for improving the overall management and quality of project information. The solution therefore is for systems that manage information and also make the information more understandable which will result in fewer errors being made. The EDMS implemented within these particular project reviews enhanced not only the collaborative culture of the project but also has the potential to enhance the collaborative culture of the industry which will ultimately lead to a more competitive and effective construction industry.

The following sub-chapters will review the individual business areas where specific applications were implemented in a range of organisations as can be seen in Table 9.4.

Table 9.4 Range and size of organisations

Cases	Company Name / organisation type	Size of Organisation / firm	Nature of Business
Project review 1	Main Contractor (consortium)	Large firm	Main contractors, civil engineering contracts
Project review 2	Main Contractor	Large firm	Main contractors, all projects
Project review 3	Main Contractor	Large firm	Main contractors, all projects
Case 4	Inspector Home Ltd	SME (small)	Independent Home inspections
Case 5	Public service	Large firm	Administration
Case 6	Main Contractor	Large firm	Main contractors, all projects
Case 7	Taylor Woodrow	Large firm	Main contractors, house builders
Case 8	CITA	SME	IT facilitator
Case 9	CIOB	SME (medium)	Professional body administration
Case 10	pH Europe	SME (small)	Logistics
Case 11	G2 Business Services	SME (small)	Performance Management

9.3 IMPLEMENTATION OF BUSINESS AND CUSTOMER CARE TRACKING SYSTEM: INSPECTOR HOME LTD (CASE 4)

Background to the business:
Inspector Home Ltd (who were established in 2001) are the UK's largest, and arguably, most well known independent new house inspection organisation. They carry out 'snagging' inspections and report on the defects found in a range of new homes for the prospective homebuyer. Operations are currently focused within the UK although the company has recently expanded into mainland Spain.

Currently Inspector Home have a staff complement hovering around 50, these being 5 office staff and over 45 regional inspectors, located around defined geographical areas within the UK.

The problem:
The problem Inspector Home faced during the early part of their existence was a common industry wide problem: relying on paper as the principle method for collecting, processing, transferring, storing and disseminating information.

With growth being a paramount issue for the fledgling business, new accounts proved difficult to attract and manage since no easy system was in place that tracked and managed customer enquiries and bookings for such a specialised business. The co-ordination of customer files and diaries was paper based with the responsibility being placed on office personnel for the creation and filing of relevant information. This traditional method of documentation control was further exacerbated when each of the 45 inspectors were independently responsible for generating, collating and transferring paper based information, but were not responsible for the subsequent filing and customer/interested party dissemination of this information. The drawbacks of this paper based system soon became apparent and the numerous issues raised included:

- Little co-ordination as a team (no sharing).
- Lack of filing index system (lack of accessibility).
- Information overload (the sheer volume of the paper based information).
- Duplication of information (tasks done repeatedly).
- Poor quality of information (what was present was difficult to mine).
- Unstructured information (the content was as entered).
- No audit trail (no system in place to track and manage).
- Little accountability (ownership was unclear).
- No 24/7/52 access to information (access limited to normal office hours).
- No workflow processes (the input was bespoke to each participant).
- Lack of consistency (no sharing of best approach).
- No central repository of information (data was insular).
- Slow processing and retrieval of project information (unstructured).
- Huge emphasis on the human to file and distribute information (erratic).

The existing use of IT:
Within the traditional paper based system only the office based staff and Directors had access to current, live, information. External parties had no right of access to this information, including each of the autonomous Inspectors. The main thrust of

IT application within the company was for basic contact administration, i.e. e-mail and routine Internet positioning via what was clearly seen as being a crude website.

The solution:
To overcome this problem, the lack of external accessibility, Inspector Home analysed the business workflow and identified key areas where significant bottlenecks occurred. These were identified as being the points most likely to improve through IT assisted re-engineering and hence, provide the greatest benefit to the business. From this analysis it became clear that the use of a number of specific IT applications would be required in order that the composite problem could be alleviated.

Implementing the chosen IT:
Three IT elements were chosen for implementation: an internal network, a company website and an integrated accounts system.

The first IT tool implemented was the internal network. This allowed all information to be stored on central servers with individual access from each desktop based PC. This particular element of the implementation process was selected under a competitive tender agreement and the eventual supplier was selected as a result of value analysis of: the cost of the proposed system, the reputation of the IT supplier and the proposed implementation timescale for the system.

The second element was the company website. A basic website (images, text, contact details, case studies) existed although this clearly needed a major upgrade and image makeover if business processes were to improve. The company website delivery provider was selected under a competitive tender arrangement and the eventual chosen supplier was selected because of the cost of the proposed system, the reputation of the IT supplier and the appropriateness of the proposed implementation timescale of the system. The website was designed to meet Inspector Home's needs (customised) and included a web based payment system for easy, faster customer transactions.

The third element implemented was the company accounts system. The old paper based system was slow, cumbersome and users found it arduous to navigate: separate files had to be created when creating customer invoices etc. All of these business vital activities have been brought under control using one automatic accounting system which creates and prints invoices automatically, resulting in dramatically reduced times taken to distribute invoice information.

The implementation process (Figure 9.4) was mainly the responsibility of two important personnel within the organisation: the Commercial Director who was responsible for the initial sourcing of the IT tools, and the Office Manager. The Director was charged with the analysis and formatting of tender documentation, developing the specification design and testing and implementation of the IT tools. The Office Manager (brought into the process to ensure a solid grounding from an end-user' point of view) ensured that the new processes not only melded with the existing, but also returned the desired efficiency gains. On-going support for the implemented systems is provided externally by the suppliers of the IT tools and managed by the Commercial Director and the Office Manager.

Initial sourcing												
Tender stage													
Specification													
Design of system													
Testing of system													
Implementation													
Ongoing support	Carried out in-house											
Months	1	2	3	4	5	6	7	8	9	10	11	12

Figure 9.4 Implementation timescale of Inspector Home's 3 IT systems

After the implementation was completed, there was an immediate impact upon the business and changes and improvements were seen in the following areas:

- Work sharing – ability to view customer files from any PC. This provided time saving and increased the efficiency of all staff.
- Accountability – management were able to monitor who was undertaking which tasks.
- Ability to track-back and demonstrate that clients had been contacted if complaints arose since all data was visible on the system.
- Web site – maximised marketing opportunities and saved time & reduced the cost of lengthy sales calls.
- Electronic accounts – saved time & money by reducing expensive accountancy costs which were to be found on the paper based system.

Other benefits derived from the implementation of these new systems include:

- Sharing of all company information and customer files (electronically).
- Ease of booking new work into the company diaries.
- Ability to look at clients files on a screen – no reliance on paper.
- No need to search file cabinets for paper based information.
- No duplication of client information: more effective communication.
- Time savings for customer care through data management.
- Professional approach with clients through speed of transactions.
- More transactions carried out electronically resulting in cost benefits in time, paper and postage although not quantifiable at the present time.

Despite the significant benefits derived from the implementation of the updated IT systems it is also important to learn from the implementation experience and note the processes that could be improved still further, such as:

- Communication with external inspectors.
- The integration of the website payment system with the accounts system.
- The use of mobile IT for remote data capture and transfer.
- Performance management of statistical data for maintenance management.

Lessons learnt from this implementation:
The most important aspect of the whole implementation process was that it enhanced the company's image through making them appear more professional in their approach. Positive feedback both internally and externally was received from a range of stakeholders. Part of the whole process was to learn, to find out the things that could have been done better or would have been done differently if the process was to be repeated.

There are concerns within the company with regards to data back-up. A feasibility study to store information and data off site has begun. This procedure is an extremely important issue because any breakdown in existing systems could result in the loss of business critical data and with it years of information. With all IT implementations it is the initial budget that determines what IT can be implemented and in this case the system was implemented and managed with the support of senior management who drove the whole process forward.

9.4 IMPLEMENTATION OF FACILITIES MANAGEMENT SYSTEMS WITHIN A PUBLIC SECTOR BUILDING (CASE 5)

Background to the business:
The Scottish building in which the IT systems implemented is a high profile one with significant architectural merit. Upon completion of the building its owners were faced with the problem of how to maintain this 'landmark' building and ensure ongoing public appreciation of the building and its users and service providers. To enable this management, four distinct IT systems were sourced and implemented.

The problem:
The maintenance and upkeep of such a high profile building was viewed as being of paramount importance. It was therefore necessary that the IT systems put into place would aid the operation and maintenance of the building. Existing maintenance processes were paper based and undertaken in rented accommodation, prior to the move to bespoke facilities within the new building. The drawbacks of the paper based system became patently obvious at an early stage. These included:

- Little integration between existing paper processes.
- Paper overload.
- Duplication of information.
- No 24/7/52 access to project information.
- No connectivity between paper and existing IT tools.
- No central repository of information.

The existing use of IT:
The existing use of IT was confined to applications located on individual PC's. No external access was possible and 95% of total IT usage was for e-mail purposes. Basic administration and accounting management consumed the remaining 5% of time.

Implementing the chosen IT:
When the move to the new building, and the Facilities Managers' bespoke accommodation, was undertaken the Facilities Management team began to use 4 new IT applications which had recently been procured. The implemented IT software packages were:

1. Helpdesk (an off the shelf product), this system deals with all facilities management enquiries,
2. A Maintenance Management System (MMS, again, an off the shelf product), this system manages the general maintenance of the building),
3. A Lighting Management System (LMS, which was a customised product); and,
4. A Building Management System (BMS, which was customised).

Three of the above systems (1, 2, and 3) were implemented by the main building contractor during the latter stages of construction operations on the project. The fourth system (the Building Management System) was implemented by the Facilities Management team. The IT provider for this system was chosen through a competitive tender arrangement and the eventual winning supplier was chosen because of their reputation as a systems supplier, the system functionality proposed, proposed timescale of implementation and most importantly the cost of the new system.

The four systems are used on a daily basis by an average of 20 people within the Facilities Management department. However, the new systems do not integrate with each other or indeed any other IT systems. This has meant that duplication of effort still occurs and there is a distinct lack of reporting and analysis features. These are carried out manually even though they patently could be executed automatically. External contractors are able to gain limited access the BMS, although any access is restricted to that available at the site itself since there is no access to any of the systems from out-with the building.

The procurement and implementation of the IT systems was the responsibility of the main contractor who was responsible for over 95% of the input to what was perceived as the user-interface of the system. The end-user was responsible for only 5% of input into user-interface design. The BMS has been constantly 'tweaked' to meet the needs of the Facilities Management team hence, the 12 month implementation timescale highlighted in Figure 9.5. A maintenance contract to maintain the IT system is in place with the IT supplier. The tweaking of the system has been carried out by a member of this IT supply team, based on site for the length of the implementation programme. The facilities management team are still in the early stage of developing new processes and this will involve further tweaking of the IT systems. The immediate impact and benefits of the new IT systems were seen in the following areas:

- No duplication of information.
- The system allowed for remote monitoring of building functionality.
- The system identified faults and areas of concern.

Initial sourcing												
Tender stage												
Specification	Stages 1 to 4 undertaken out-with the facilities management department											
Design of system												
Testing of system												
Implementation												
Ongoing support												
Months	1	2	3	4	5	6	7	8	9	10	11	12

Figure 9.5 Implementation timescale of the BMS facilities management IT system

Other benefits that derived from system usage were:

- Less reliance upon paper.
- A central repository of information.
- An effective maintenance management programme.

Despite the significant benefits that were experienced from the use of the new systems it is also important to learn from the experience and examine the issues that could be improved further such as:

- Less reliance upon customised systems.
- The integration of the IT systems to be able to produce reporting documentation.

Lessons learnt from this implementation:

The implementation of the new IT systems was very favourable from a human perspective because the systems provided a significant amount of improved information that influenced the decision making process and final decisions. There were and still are concerns however about the reliability of such information and if the implementation was undertaken again the end-users would like more control over the implementation in order that they might better inform decisions and be able to make better decisions.

From a process point of view the implementation is still on-going and there are still some minor issues to be resolved regarding the user-interface of the system. In essence the technological problems could have been alleviated if the people who use the system had been involved in the implementation at an earlier period. The technology implemented has helped the Facilities Management team deliver an enhanced service, but with the system not being fully functional, it means the team are not achieving the full potential of the system and thus losing out on additional benefits they should be accruing.

The principal lesson to be gleaned from this particular implementation is that the thoughts and opinions of the end-user should never be ignored and they should be brought into the implementation process at the earliest possible stage in order to maximise the potential of the IT systems.

9.5 IMPLEMENTATION OF AN EDMS FOR THE CONTROL OF PROJECT DOCUMENTATION (CASE 6)

Background to the business:
This particular construction contractor has been established for over 130 years and are a leading UK building and civil engineering contractor with wide experience on major projects covering all of the major procurement routes. The main strength of the business is its capability to manage a wide range of projects involving multi-disciplinary teams across a wide geographical spectrum and they achieve this through a culture of co-operation with clients and construction professionals alike. The company strives towards the improvement of quality which includes the adoption of ISO 90001 as a quality assurance standard.

The problem:
The problem facing the organisation was the control and co-ordination of project based information. With sustained growth and expansion being of paramount importance for the business, some way of controlling project documentation had to be found which would alleviate many of the paper based problems experienced within the industry. The co-ordination of project documentation was both paper and IT based with the responsibility being placed on individual personnel for the creation and filing of relevant project information. The drawbacks of these non-integrated systems were numerous and included:

- Little co-ordination on a project basis (no sharing of information).
- Lack of filing index system (lack of accessibility).
- Information overload.
- Duplication of information (tasks repeatedly carried out).
- Poor quality of information.
- Unstructured information (the content was as entered).
- Lack of consistency (information consistency).
- Slow processing and retrieval of project information (unstructured).
- Huge emphasis on the human to file and distribute information (erratic).

The solution to the problem:
To overcome this lack of control and co-ordination of project documentation, the organisation analysed their existing business processes and identified where significant problems in the information supply chain occurred. From the analysis it became very clear that the use of an EDMS would be required in order that the problem of control and co-ordination could be overcome. This type of system was needed because the system was to be adopted on all of the organisation's construction projects and to date the system has co-ordinated documentation on over 200 major construction projects.

The system itself is a Lotus Notes application and has notable features such as its audit trail, lockdown features for documentation, document access history and document notification process. The EDMS includes a correspondence database for handling day-to-day construction correspondence, a technical documents database which handles all technical queries, variations, instructions and request for instructions and a database which logs all project documents such as drawings, specifications and method statements. The aim of the EDMS was to create a robust system that could be used for document/information management on all of the organisation's projects.

Implementing the chosen IT:
The chosen IT system was a customised system designed by the EDMS manager within the organisation and the eventual IT supplier was chosen because of the cost of the proposed system, the company's technological ability with the software to be adopted, the reputation of the IT supplier and the implementation timescale which can be seen below in Figure 9.6. A technical contract is in place with the IT supplier that allows for continuous upgrading of the system.

Initial sourcing												
Supplier discussions												
Specification												
Design of system												
Development and testing												
Pilot the system												
Implementation	Roll out to each new project after pilot											
Ongoing support												
Software alterations										On-going		
Months	1	2	3	4	5	6	7	8	9	10	11	12

Figure 9.6 Implementation timescale of the EDMS

The implementation of the new IT system was mainly the responsibility of the EDMS manager. The EDMS manager together with an internal colleague and the IT supplier built the IT specification and the EDMS manager was also responsible for the testing and piloting the new system. The EDMS manager as an end user defined the end user interface. Only after the conclusion of the pilot schemes did the other end users within the company have any influence on the design of the user interface. On-going maintenance is provided by the IT supplier and support and training for the implemented system is provided internally by the EDMS manager and a number of regional business systems managers located throughout the UK. The EDMS was piloted at four regional offices located within the UK over a 3 month period. After the successful pilot period the EDMS was rolled out into every new project that came on stream.

Over 1200 personnel within the company have access to the system although this is based upon a hierarchical structure depending on the role of that person within the organisation. In other words 'on a need to know basis' (see Figure 9.7). External disciplines such as sub-contractors do not have access to the system although clients, joint venture organisations and members of the design team do have access on a project specific basis although the documents they can view are technical based documents (change management and drawing information).

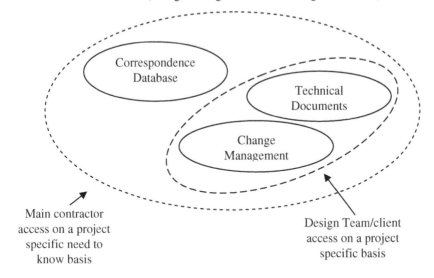

Figure 9.7 Access to the EDMS depending upon project role

After the implementation was completed, there was an immediate impact upon the organisation and changes and improvements were seen in the following areas:

- Sharing of all project documentation internally on a need to know basis.
- No need to search file cabinets for paper based information resulting in time savings.
- Less duplication of project documentation.
- More effective external communication.
- More transactions carried out electronically resulting in time, cost and quality benefits.

Other benefits derived from the implementation of these new systems include:

- Accountability – management were able to monitor who was undertaking which tasks.
- Project data backed up on and off site according to project set-up.
- Constant upgrading of changing working procedures.

Despite the significant benefits derived from the implementation of the updated EDMS system it is also important to learn from the whole implementation experience, and note the things that may have been done differently should the same implementation be carried out today. There are no doubt always a range of areas that could be tightened up and others that should be lessened: what is important and useful is to consider the broad approach. Example areas include:

- The control of e-mail.
- Compulsory training of staff.
- Ensure sufficient key support staff available.

The current working environment:
The distribution of information from within the organisation is currently handled by the EDMS although the amount of documentation created on a particular project depends upon the type of project it is, the working environment, how many external parties are involved and also the procurement route adopted. For example, on a £10m project, the amount of documentation created would be in the region of:

- Correspondence documents – 25,000
- Change management documents – 5,000
- Technical documents – 13,000 giving a total of 43,000 documents

Many projects undertaken in today's current construction environment are larger in value. The organisation has also indicated that a £100M+ contract which is only a fifth of the way through the project programme has already created:

- Correspondence documents – 129,000
- Change management documents – 18,000
- Technical documents – 18,000 giving a staggering total of 165,000

The above figures demonstrate the extremes of information management within the construction industry. The findings from this particular case review demonstrate the benefits of adopting EDMS as a tool for controlling and co-ordinating project information. Information and documentation created on this scale however requires skilled personnel who understand the working ways of construction and who are able to design operate and upgrade systems in order to handle the massive amounts of information currently being created on construction projects.

9.6 IMPLEMENTATION OF A DIGITAL PEN AND PAPER SNAGGING APPLICATION: TAYLOR WOODROW (CASE STUDY MATERIAL COURTESY OF DR SARAH BOWDEN, ARUP AND PAUL GOODING, BSRIA - COMIT WWW.COMITPROJECT.ORG.UK) (CASE 7)

Background to the business:
Taylor Woodrow are a leading UK contractor and major house builder (Bryant Homes). Like many other major contractor and house builder, Taylor Woodrow has become aware that they are spending too much time 'snagging'. Snagging or the rectification of defects is a common industry wide problem and is estimated to account for the last 1% of a project. The rectification of snags is a particularly critical phase of the project given that it comes close to the final handover stage of a project. On larger sites snagging can become a laborious and time consuming task and one which requires a great deal of patience for the person who has been designated as the snagging manager. This particular case study discusses the implementation of a digital pen and paper based snagging system on a major project in Scotland.

The snagging problem:
The problem the contractor was facing on this particular project was the capture and processing of snagging information. Recording snagging data using traditional paper methods was proving time consuming which resulted in sub-contractors receiving snagging data many days after the items were originally identified. This delay in disseminating snagging information is one of the main causes of delays to projects and can be very costly. The original snagging process consisted of the following steps:

1. Snags listed on A4 paper with necessary sub-contractor information noted.
2. Snagging information re-typed into the computer system back in the office.
3. After data interpretation, the snags were passed back for final checking.
4. List of snagging items handed out to sub-contractors.
5. Sub-contractors receive snagging items.
6. Sub-contractor actions the items and notifies the site manager.
7. Site manager undertakes further snagging inspection.

The solution to the problem:
To overcome this lack of control and co-ordination of snagging documentation, Taylor Woodrow chose a revolutionary digital pen and paper based system which allowed the site operatives collecting the data to continue collecting data in a familiar manner with the added bonus that this data was being automatically processed into a web-based system using hand writing recognition software. The technology can be adopted and data sent directly via mobile phones or docking stations attached to a user's PC. Using this system, the sub-contractors are in almost immediate receipt of snagging items. Taylor Woodrow and the IT supplier Sysnet re-designed and re-formatted the off-the-shelf solution's digital forms to allow the recording of project specific information although 'Snagmaster™' can be adopted

with limited customisation on a wide range of projects. The previous working method involved the re-typing of hand-written information into an electronic format. Using the digital pen and paper solution allows snags to be identified on paper, immediately recorded, checked, actioned, and signed off within the system. The whole process is carried out within a secure environment with controlled access to the database in accordance with what role the snagging party plays within the project. All snagging data can be interrogated, reports produced and snags issued electronically. The main benefits of this system are:

- Improved confidence in snagging data.
- Speed of the snagging process has been dramatically improved.
- Quicker data interrogation.
- Fully searchable system.
- Quicker dissemination of snagging information.
- Built in access features (you only see what you need to see) can be provided to a number of project parties.
- Complete electronic records of snagging data from a particular project.
- Secure electronic environment with the confidence of knowing data has been issued where it should be issued.
- A familiar and comfortable working method is still used although all paper based information becomes electronic.
- Electronic images of paper based information are stored within the system.

The cost of the new digital pen and paper system for this particular project was circa £4,500 and, including running costs the total cost for the first year was just over £10,000. The key areas of expenditure were in the design and customisation of the forms, printing of the snagging forms and the associated software licences. With the new system operational, the time savings were immediate and resulted in one fewer staff member required to process snagging data. This time reduction alone within the first year has been estimated to save over £20,000 per annum and there are also great benefits and savings in general administration of the snagging process. The new snagging process consists of the following steps:

- Snags identified on digital pen and paper forms completed on site by a member of the project team (the pen can hold up to 500 individual snagging entries in its memory).
- Snags identifier returns to office and synchronises snagging data into web based database (at this stage the software converts the handwriting and records the time and signatures of the identifier. An image of the form is also kept within the system).
- Snags are checked and verified by the site manager or site administration personnel.
- Snags issued electronically through the database to the sub-contractors.
- Snags are rectified by the subcontractor.
- Sub-contractor notifies progress to the site manager using the digital forms or the action features within the database.

- Site manager inspects the snagging items and signs these off using a third form which automatically moves the snagging items into a 'complete' view within the database.

Implementing the chosen IT:

The implementation of the chosen digital pen and paper based solution on this project was the responsibility of the site manager, company IT staff and the external supplier's IT staff who formed a very close working relationship. The companies involved had previously worked together implementing other IT solutions. The application, which was the first time this particular application was adopted within the UK, was implemented onto the project within six months as demonstrated in Figure 9.8. The immediately familiar pen and paper user interface meant training was very simple, only written process instructions were provided to the user.

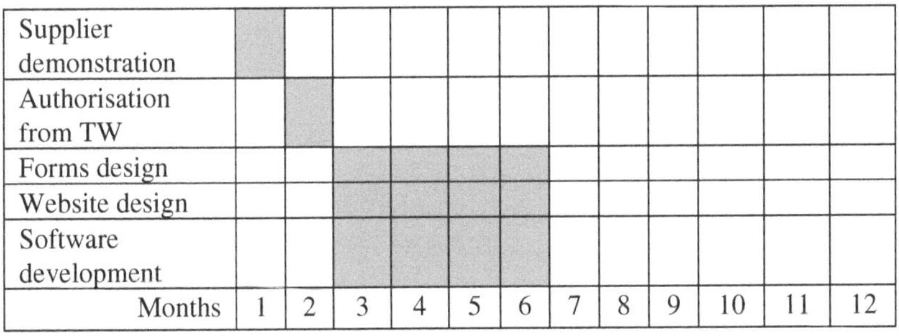

Supplier demonstration												
Authorisation from TW												
Forms design												
Website design												
Software development												
Months	1	2	3	4	5	6	7	8	9	10	11	12

Figure 9.8 Implementation timescale of the digital pen and paper based snagging solution (estimated)

Business benefits tend to be centred on time, cost and quality. This particular implementation was no different. The costs saved in this project are largely due to the reduction in time it takes to input data and also that of reporting the data. Time is saved whenever data is required. The forms are so intuitive that very little human input is required when the information is populated into the database. Printing and postage costs associated with the dissemination of snagging data are almost eliminated because the snagging transactions are carried out by e-mail. This makes the reporting and rectification of the snagging items much faster. From a quality point of view, the contractors are more inclined to complete the job. The database is so user friendly that the contractors are less likely to miss any snagging items applicable to them which, improves accountability and traceability.

Despite the significant benefits, it is also important to note the lessons learnt from the implementation process. During the project, the construction team learnt that the handwriting recognition software was able to interpret clear writing and capital letters more easily. A drawback of the software is that it encounters problems when trying to interpret unclear writing. The new technology also prompted the project team to improve the structure of the forms which helped the operatives to complete the forms more easily.

Evaluation of the system is taking place and Taylor Woodrow may extend the use of the technology to other projects. The great thing about the whole system is that the technology behind the digital pen and paper application is almost hidden which has greatly reduced the common human barrier, that of resistance to change.

To view the Taylor Woodrow case study in full and other mobile IT construction case studies please visit www.comitproject.org.uk

9.7 CONSTRUCTION INFORMATION TECHNOLOGY ALLIANCE (CITA): PILOT PROJECT FOR AN ELECTRONIC PURCHASING SYSTEM (CASE 8)

Background to the business:
The Construction IT Alliance (CITA) was formed in 2001 by two of Ireland's prominent academic figures in the area of Construction Information Technology: Alan Hore of the Dublin Institute of Technology and Ken Thomas of the Waterford Institute of Technology, along with key construction industry professionals who had a vision; that their group would assist in harnessing the potential of information and communication technologies for the Irish construction sector. The key driver in the development of this organisation was the need to increase the extent of IT knowledge and to take full advantage of current and emerging IT technologies. The main benefits of being a member of CITA include being part of a wide community of construction industry professionals who are determined to work together to realise the benefits that IT can bring to their professional lives.

The problem:
The traditional process of purchasing materials involves a mainly paper based process although some degree of IT was used. Typical documentation created includes purchase orders, delivery dockets and supplier invoices. The existing IT utilised was standalone software solutions which were accessed by over 100 people which included purchasers, site administrators and accounts personnel. No external access was provided as the system was in-house only. The main use of IT in house was for administration, materials ordering and materials delivery. However, there are very poor levels of integration between and within the respective departments and the software and paper based solutions did not integrate with each other in any way. The drawbacks of the mainly paper based system included:

- Reliance upon paper documentation (too much paper).
- Duplication of information (tasks repeatedly carried out).
- No central holding repository for information.
- Lack of consistency (information consistency).
- Slow processing and retrieval of purchasing information (unstructured).
- Errors in transcribing paper documents.
- Mislaid and lost documents.
- No formal filing system.
- Extensive re-keying of paper based information into IT systems.
- Huge emphasis on the human to file and distribute information (erratic).

The solution to the problem:
The overall aim of the pilot project is to re-engineer the purchasing process within a contractor's organisation, by enabling an electronic three-way match of the purchase order, delivery docket and invoice data, thus enabling an improvement in both productivity and overall administration costs per transaction. The proposed ICT infrastructure (web based system) to be adopted will involve the electronic transfer of purchase order, delivery notes and invoices via a central web-based repository known as the COINS ETC HUB. The HUB is able to translate any

incoming EDI, XML or spreadsheet documents from either the contractor's COINS ERP system or the supplier Unison ERP system. The HUB is also capable of sending outbound messages in these formats for basic information purposes or integration into other customer's ERP systems.

The operation of the proposed solution will involve the following sequential tasks:

1. The contractor creates an Open Order in COINS. In advance of this communication, the contractors will have negotiated a schedule of prices for particular products from the supplier.
2. This creates an automatic fax to site providing details of the purchase order number.
3. The site personnel call-off the material by telephone, fax or email to the supplier.
4. The supplier in turn creates a purchase order on their supply order desk.
5. This creates an ePO, which is electronically sent to the COINS ETC HUB. The ETC Hub will convert the data into a BASDA XML message, which in turn is forwarded to the contractor's back-end database and populates a line item on the contractor's purchasing workbench.
6. The ePO created by the supplier is dispatched to the O2 Instant repository, which in turn routes the message to an appropriate handheld computer.
7. The supplier delivers the material to site and the contractor electronically signs the PDA. The ePOD is routed back to the O2 Instant repository.
8. The ePOD is routed to the COINS ETC HUB.
9. The ePOD is routed to both the contractor's and the supplier's back-end database and populates line items in their respective ICT system, thus created an eGRN.
10. The receipt of the ePOD in the suppliers back-end system, will allow the supplier to create an eInvoice from ePOD and ePO data.
11. The supplier eInvoice is routed via the COINS ETC HUB to the contractor's invoice workbench on the COINS ERP software.

Implementing the chosen IT:
The IT implementation involved realising an electronic three way match of the purchase order, delivery note and supplier invoice. The chosen IT system for the pilot project was an off the shelf product and the eventual IT supplier was chosen because they were the preferred IT vendor of the contractor and because they were able to demonstrate the business benefits of using IT. The total implementation timescale can be seen in Figure 9.9.

The new pilot IT system does integrate with other IT systems. For example, the web based HUB integrates with the ERP system of both the client and contractor and the electronic proof of delivery also integrates by use of agreed XML messaging formats. External parties cannot however gain access to the system due to security issues. The implementation costs are not clear because the project was a pilot and therefore the impact and change in business processes achieved as a result of implementation cannot be measured. The hope is to move onto a fully fledged operation which will monitor costs and benefits and report back on these. There are however some minor benefits of the new IT system which are shown in the following list:

- No scanning of paper documentation.
- Little re-keying of data in to the system.
- A reduction in the amount of paper used and paper traffic.

Initial sourcing	▓	▓										
Tender stage			▓		·							
Specification				▓								
Design of system					▓							
Testing of system						▓	▓					
Implementation								▓	▓			
Ongoing support										▓	▓	▓
Months	1	2	3	4	5	6	7	8	9	10	11	12

Figure 9.9 Implementation timescale of the pilot project

Since the pilot project began, the number of people allowed access to the system has dropped to below 50 due to efficiencies in working procedures being experienced. Users have also commented upon how easy the system was to use, how it made difficult processes more efficient and the fact that the IT system actually worked the way they wanted it to work. It is also important to reflect and discuss areas of the implementation that could have been improved. The implementation could have been handled better from a project management point of view and the pace of the implementation was too slow when considering that it was a pilot scheme. From a user point of view more training should have been provided and additional functionality should be added which would further improve the process. An objective however is not only to improve the process but re-engineer the process. Despite these efficiencies and improvement in working methods, a decision on weather to implement the IT system across the board will not be made until after completion of the pilot project.

For more information on this pilot project please visit http://www.cita.ie/

9.8 IMPLEMENTATION OF A NEW MEMBERSHIP DATABASE: THE CHARTERED INSTITUTE OF BUILDING (CIOB) (CASE 9)

Background to the business:
For almost 170 years the CIOB has been the leading professional body for managers in construction with circa 40,000 members worldwide in over 94 countries. The CIOB is a dynamic institution that works to a rolling and constantly evolving corporate plan. The CIOB leads the way in establishing, promoting and maintaining standards of excellence throughout the construction industry. The CIOB is dedicated to raising standards throughout the world to the benefit of customers, clients, contractors and its members.

The problem:
The problem facing the CIOB was how to retain existing data, allow for its upgrade and, facilitate the recording of future membership data. A bespoke system was already in place although there were many drawbacks associated with using this system including:

- Lack of flexibility around subscription billing.
- No facility for electronic storage linked to database.
- Lack of support from the software supplier.

The main issues surrounding the use of the old system (before the implementation of the new) were mainly to do with financial administration. The old system could not/was unable to:

- Do billing (invoicing) in foreign currency.
- Record fees/subscriptions/payments in a foreign currency.
- Accept and confirm receipt of payments on-line.
- Was poor in relation to document management.
- Was slow in processing and retrieving membership information (unstructured).
- Was not accessible to Regional and overseas CIOB administrators (externals).

Twenty internal members of the membership team had access to the old system, externals were denied access because the technology was not in place that would allow them to do so. The main use of the old members IT system was for records management, accounting and administration. A system was required that allowed the membership team to bring together all of the administration and invoicing tasks into one database and make this database accessible to all.

The solution to the problem:
The new IT system chosen was an Oracle based off the shelf product. The system supplier was chosen under a competitive tender arrangement and were eventually accepted due to their reputation, the system functionality, cost of the system and the proposed implementation timescale of the new system. The new system integrates seamlessly with the Internet and the organisation's website. Over 100 users now

have access to the new system, including all members of the membership team, members of the education library and more importantly, regional staff. By allowing external access through a secure log in, members across the world can now view their membership data online via the web access module and update accordingly.

Implementing the chosen IT:

The chosen IT system was implemented over a 12 month timescale period. The decision to implement the new system was discussed at the CIOB's Policy and Finance Committee and passed to its Council for ratification. A fully supportive decision taken by senior management within the CIOB resulted in the new system being jointly customised and designed, to some extent, by members of the IT supplier's team and also end users within the CIOB. The investigation and feasibility costs have been kept low by undertaking the brief and feasibility study in-house. The key activities within the implementation were as follows (a timescale can be seen in Figure 9.10)

The feasibility and initial investigation was undertaken by a person within the membership team (an end user) who examined the use of similar systems in similar organisations. From this initial investigation they settled on Concept (name of system) from a shortlist of 5. (No cost to the organisation).

1. The system brief was written with end user input. Staff needs were identified with the final brief being written by a staff member. The cost of this stage is estimated at circa £2,000.
2. Software functionality – minor customisation was required; cost £4,000.
3. Communications and data storage were included within the £4,000 above.
4. Consultancy was arranged with an external supplier who read over the tender document and brief; cost £1,000.
5. Installation of the system was undertaken by CIOB IT staff; cost £5,000.
6. Staff training: 2 key staff were trained and then the trained became the trainers, although a drawback to this was the time taken to train: in excess of a solid one week; cost £7,000 (costs have been kept low by training staff to train)

Initial sourcing												
Initial Sourcing												
Tender Stage												
Specification												
Testing of System												
Implementation												
Ongoing support	Stage undertaken as necessary											
Months	1	2	3	4	5	6	7	8	9	10	11	12

Figure 9.10 Implementation timescale of membership database within the CIOB

After the implementation of the membership database was complete, there was an immediate impact upon the organisation as a result of the implementation.

The major changes for the business arise in the operation of the underlying membership processes:

- Capture of contact details and correspondence of members.
- The on-line facilities available.
- The ability to bulk e-mail members.
- Less duplication of project documentation.
- More transactions carried out electronically resulting in time, cost and quality benefits.

The changes have enabled the organisation to deliver what is perceived as a 'slicker' process to its members and has also resulted in significant savings with regards to postage, time and staff costs. More knowledge on members is also held within the central repository and is readily accessible. Other benefits derived from the implementation of the new systems include:

- The ability to bill/invoice in sterling and foreign currency.
- Record all correspondence.
- Record qualifications held by members.
- Automation of business processes.
- Payment of subscriptions on-line.
- Printing of payment receipts from web-site by members.
- Complete history of grade changes.
- Organisation maintenance – companies, teaching institutions etc.
- Committee details.

The implementation of the new IT systems was extremely favourable from a human perspective because it helped to pull the membership team together. A drawback of the successful implementation from a people point of view however was the levels of stress it caused within the implementation team.

From a process point of view the implementation made the membership team look more closely into current business processes and put measures in place on how they would improve them. If the implementation was to be started all over again however the team have indicated that the implementation should be started earlier so there are longer training periods towards the end of the process. The technology implemented has helped not only the membership team but has also aided members worldwide who now have access to a range of data and resources within the CIOB.

9.9 PH EUROPE LTD: IMPLEMENTATION OF AN ASSET TRACKING SYSTEM USING RADIO FREQUENCY IDENTIFICATION DEVICES (RFID) (CASE 10)

Background to the business:
pH Europe are a container (stillage) supply company focussed on delivering an international sales and rental service. The company, based in Huddersfield, are organised into 3 distinct areas: stainless steel containers, plastic containers and environmental storage products. Each of the stillage units (containers) were supported by a comprehensive, paper based, fleet maintenance and tracking system. Extensive storage, repacking and cleaning arrangements further enhanced the service.

The problem:
The problem facing pH Europe was that of how to track the assets belonging to them and the pH Europe rental fleet. Before the implementation of the new tracking system, no IT system was in place that aided the tracking process. The main use of IT within the company was limited to basic document processing. IT was mainly used for basic administration with a significant reliance upon paper based documentation. The drawbacks of this system were clearly apparent and included:

- Assets (containers) becoming lost.
- Limited electronic storage of information.
- No way of knowing where the asset is with regards to location.
- Sums of finance written off due to the lack of a tracking system.
- Poor performance in relation to document management.
- Poor in relation to asset management.
- Slow in processing and retrieving asset information.

A system was required that allowed the company to track their assets. Containers were being lost, both in transit and at client depots, and this was proving a significant drain of the company's finances. Substantial sums of finance were budgeted for replacement of the assets.

The solution to the problem:
The new IT system selected is an Intranet based system which was selected under a single stage tender process. The IT supplier was chosen because of their expertise, reputation, the system functionality offered, cost and more importantly, the proposed timescale of the implementation. The new system (finder^{TM}) provides a remote container tracking service that requires no operator intervention, allowing virtually real-time reporting on container stocks and movements to and from your customers and suppliers. The new system is heavily customised although it does not integrate with other IT systems within the business. The system was designed under a true 50/50 collaboration by the end user and the IT supplier, and only 2 days training was required in order to fully commission the system and have it operational. Currently only 10 members of staff have access to the new system and these are mainly at management level. External parties are able to access the system although this tends to be on an administration and support basis only.

Implementing the chosen IT:

The chosen IT system was implemented over a 12 month timescale period. The decision to implement the new system was fully supported by senior management within the company. The key activities within the implementation were as follows (a timescale can be seen in Figure 9.11).

1. Initial sourcing: the feasibility and initial investigation was undertaken by a nominated person within the Product Support Team (an end-user) who carried out an extensive survey of potential systems and examined the use of these similar systems in similar organisations. From this initial investigation they settled on a concept from a shortlist of 3. (The cost for this stage of the programme was underwritten by the company securing funding for a Knowledge Transfer Partnership, KTP).
2. Tender stage: the system brief was written with direct end-user input. Staff needs were identified with the final brief being written by a staff member. The cost of this stage was rolled into the KTP.
3. Design of the system: specification, software functionality, heavy customisation was required. The cost of this stage was rolled into the KTP.
4. Communications and data storage were included within the stage above.
5. Installation of the system was undertaken by pH Europe staff and the IT supplier. The cost of this stage was rolled into the KTP.
6. Staff training: all staff that has access to the system was provided with two days training. The cost of this stage was rolled into the KTP.

Initial sourcing	▓											
Tender Stage			▓	▓								
Specification	▓											
Design of System		▓			▓	▓						
Testing of System								▓	▓			
Implementation										▓		
Ongoing support					▓	▓	▓	▓	▓	▓	▓	▓
Months	1	2	3	4	5	6	7	8	9	10	11	12

Figure 9.11 Implementation timescale of the asset tracking system

After the implementation of the system immediate changes in business processes were experienced by the company. These include:

- The automation of business processes.
- Reduced labour intensity of the processes related to assets.
- Complete electronic record of existing assets.
- Improved stock taking process.
- Improved scheduling process.
- Less duplication of project documentation.
- Improved maintenance procedures.

The Newly implemented system basically helps the company to manage its fleet of assets efficiently and it provides a better service to our customers as well as enabling us the company to manage our customer relationships in an open and harmonious manner. The developed system was clearly novel, and as such the company utilised its novelty to enter for, and gain, a number of prestigious business awards. A few of these awards are listed below to illustrate the recognition afforded the new system.

1. The IM2004 Information Management Awards, Supply Chain Management Category.
2. Computerworld magazine's Mobile & Wireless Best Practices Award for Deploying Wireless Mobility in the Enterprise Westin Kierland Resort, Scottsdale, Arizona, June 13, 2005.
3. E-Commerce Awards – Yorkshire and Humber Regional Final, 'Best Use of Mobile and Wireless Technology' Category.

9.10 IMPLEMENTATION OF A PERFORMANCE MANAGEMENT SYSTEM: G2 BUSINESS ASSOCIATES (CASE 11)

Background to the business:

G2 Business Services based in Glasgow, UK have devoted thousands of hours of research and development to come up with a solution to the challenge of managing service performance across multiple functions and multiple sites within various industrial sectors. In an operating environment where change is the only constant, control is crucial, and g2BS's pioneering performance management software application is a positive response to business needs and expectations.

The culmination of the thousands of hours of research is an information management portal that captures key data, turns it into information and presents that information as meaningful business knowledge. g2-Metrix® (the IT system) enables you to manage not only your own business performance but the performance of your contracts, contractors, sub-contractors and suppliers. It is a performance management tool with multiple functionality features that allow an organisation to pinpoint areas of concern. g2-Metrix® is a secure web-enabled application that is customised to help you measure against key performance indicators and key performance outcomes. The advantage for clients and suppliers alike is that they can share real-time, historic and benchmark information on a 24×7×365 basis.

The problem:

The problem facing many organisations is how to manage performance. The problem that g2 faced was that the existing practice of managing performance was carried out using software tools (basic spreadsheets) that were inappropriate for managing performance which was causing massive administration and reporting problems. The problem is further exacerbated by the fact that over 100 people had access to the system which included facilities managers and performance managers although externals could not access the old system due to security problems. Other IT systems also existed within the organisation (helpdesk) but there was no integration between these systems which further added to the administration problems. Basically a disparate and fragmented approach to performance management was in operation because no IT tools existed that could have changed this fragmented process. The drawbacks of the spreadsheet based system were numerous and included:

- Fragmented performance management information.
- Heavy burden upon administration staff.
- Heavily resource intensive.
- Delays in key decision making.
- Limited component correlation.
- System incapable of managing performance in multi team environment.
- Lack of reporting software.
- Poor performance in relation to document management.
- Slow in processing and retrieving performance management.
- Reliance upon the individual to analyse data.

Technology was required that would take care of data collection and more importantly the reporting process. The system could then be rolled out throughout industry and outsourced into the supply chain. Human resources used before the implementation of the new IT system could then be channelled into the analysis and reporting of performance management data.

The solution to the problem:

The new performance management system is a web based system with the IT supplier being chosen by the in house management team. The system (g2-Metrix®) is a heavily customised product which was 100% designed by the management team and the end users. The system also integrates with other systems within the business such as the helpdesk, e-mail and other databases. Over 100 people including facilities manager's suppliers and database administrators now have access to the new system and this includes external clients who were previously unable to view their performance related data. Initial staff training was heavy and was undertaken over a 5 day period. The system works on a secure log in basis providing full security features and is upgraded on a regular basis to take into account user feedback. In operation g2bs have replaced this chaotic spreadsheet approach with a system of greater Efficiency. This system is designed for multi-services/locations and multiple suppliers/teams and provides time for performance analysts to analyse the data. Crucially the system gets the correct information to the correct people when it is needed allowing appropriate and fast decisions to be made. G2 Metrix allows many users to readily access this information using exception, benchmark, trend, analytical and report functions.

Areas for improvement can be found immediately, aided by the drilldown facility within the system, and acted upon before it's too late. Based on current ongoing research the information is displayed using the popular traffic light system. An organisation can also include its customers or internal people in measures and reports allowing the organisation to obtain a holistic view of company performance. This information is measured and collected on a simple met/failed basis with scores and weightings building in tolerance. The clustered groups of KPI's are channelled in to KPOs and are derived directly from services offered and the according level of service to be provided. Some of the most significant benefits gained from g2 Metrix are shown in Table 9.5.

Table 9.5 Benefits derived from adopting g2 Metrix

Substantial reduction in time and cost	Administrative burden removed
Allows resources to focus on performance not data collection	Provides real time information to make decisions
Transparency	Trend, benchmark and compare data
Correlation of different components	Reduction of risk
Ability to manage exceptions	Maximisation of efficiency and effectiveness

Implementing the chosen IT:

The system supported by senior management was implemented over a 12 month period as shown in Figure 9.12. The cost of the solution and the key activities can be viewed in the Table 9.6.

Table 9.6 Key activities within the implementation

Description	Key activities/actions	Cost (£K)	Outcomes
Feasibility/investigation costs	Mathematical modelling	6	Initial proof of concept
The IT system brief	Platform / interface	5	Development specification
The software functionality	Functional spec/use cases/business logic/workflow	12	Database design, user interfaces
Communications infrastructure	Interface with other applications	9	Middleware
Data storage and retrieval	Databases design, languages	27	Operational database development, end user presentation
Installation of system	Deployment to specific environments	5	Operational platform
Staff training	Systems, functions, customisation	5	Knowledge transfer
Additional staff resources	Administration, web design team	30 (p.a)	Operational administration
Additional support	Hosting	12 (p.a)	Internet infrastructure
Other resources	Outsourced maintenance	7.5 (p.a)	Server hardware, additional software
	Total implementation	69	
	Total running costs	49.5	

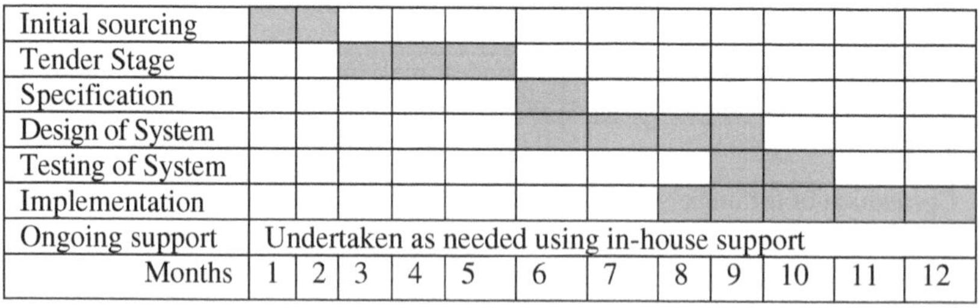

Initial sourcing												
Tender Stage												
Specification												
Design of System												
Testing of System												
Implementation												
Ongoing support	Undertaken as needed using in-house support											
Months	1	2	3	4	5	6	7	8	9	10	11	12

Figure 9.12 Implementation timescale of the performance management system

After the implementation of the performance management system immediate changes in business processes were experienced by the company. These include:

- 15 stage performance management process being reduced to 2 stages.
- Resources focused on value analysis.
- Cohesive web environment created to enable strategic performance management.
- Decision making made easier.
- A Return on Investment (ROI) of 4:1 against traditional practice.
- Communication – improved team dynamics and motivation.
- Realistic data analysis and data collection carried out automatically.

Despite the benefits it is important to learn what was also good and bad about the implementation process. From a people point of view the implementation was good because it saved time, enhanced decision making and made better use of resources. The poor aspect of the implementation was the lack of market knowledge. From a process point of view, the ROI was an immediate benefit. However the end users needed educating which involved using up valuable time and resources. The company have also indicated that they would re-examine the bigger picture within the performance management environment. From a technology point of view the online presentation features require further upgrading. A negative aspect of the implementation is that client IT infrastructure is often inadequate and the use of different IT platforms and infrastructure should be considered.

The development of g2-Metrix® and other associated software was driven by the company and practitioners alike to meet a market gap identified by ensuring that end user knowledge and existing performance management procedures were well understood and documented. The new system enables the company to work with a range of partners in a wide variety of business sectors who all work to one aim – that of managing performance to improve business processes and create and maintain competitive advantage.

9.11 SUMMARY

The case reviews have shown that the organisations, irrespective of their size and nature, have benefited from the implementation of the IT. These benefits are outlined in Table 9.8. It is important to emphasise that all of the organisations did not experience all of the benefits: rather, they enjoyed benefits which were particular to their needs and expectations.

These benefits can be summarised by saying that the development of central repositories of information ensured that information became much more accessible and useful. The data actually gathered was able to be processed much more rapidly, and with automation of the data gathering, so much of the drudgery associated with this task fell away.

The enhanced data processes also brought about an improvement in the consistency of actual data. The eradication of duplicated information, and the duplication processes, proved of real worth to the businesses. The reduced emphasis on human interaction negating many of the subjective approaches. This of course had a direct impact on the administration costs, which were invariably significantly reduced.

These savings, coupled with the reduced time taken to gather, access, process distribute, and store information also brought about a perception of real gains in how day-to-day business was conducted. The direct financial savings generated, and the improved productivity, meant that the organisations saw handsome returns on the investments made. Part of these productivity improvements centred on the new, improved, structured workflows which aided the people in delivering the desired business outcomes.

The increase in the number of participants able to access the relevant systems and data also meant that the organisations saw improvements in their business processes and procedures. Having no real restriction on the number of users was clearly perceived as a boon.

There was clear support for the view that communications were generally much improved after the IT implementation. Not only more efficient, but more effective. Indeed the reduction in the reliance upon paper was seen as a significant move for the organisations. Many thought that the demise (or at least reduction) in the paper mountain would generate issues surrounding the worth and value of the electronic transactions. These fears were found to be totally unfounded.

The fact that there were now clear audit trails, with precise information on each activity and participants, meant that accountability increased. Not only did the accountability increase, but also the speed with which responses and actions were transacted improved. Dissemination of relevant information improved in the majority of the cases, with users finding the accessibility of information a real bonus when compiling and transmitting communications.

Table 9.7 Features of the implemented systems

Cases reviewed	Time taken for implementation (months)	Average Costs	Average Savings	Customised or off shelf systems	IT system support	Integration with other IT	Average Training (days)
Projects 1-3	N/A	N/A	Not stated	Customised	External	No	N/A
Case 4	5	N/A	Not stated	50/50	Internal	No	N/A
Case 5	12	N/A	Not stated	50/50	External	No	N/A
Case 6	7	N/A	Not stated	Customised	External	N/A	3
Case 7	6	£10k	£20k est.	Off shelf	External	Yes	1
Case 8	12	N/A	Not stated	Off shelf	Internal	Yes	N/A
Case 9	12	£23k	Not stated	Off shelf	Internal	Yes	5
Case 10	12	N/A	Not stated	Customised	External	No	2
Case 11	12	£69k	Not stated	Customised	Internal	Yes	5

Table 9.7 highlights the features of the various implemented systems. Approximately half the systems were customised and half were off the shelf products which, indicates that available software is appropriate for use in some situations but also a requirement for heavily customised products. What is noticeable from both Table 9.7 and Table 9.8 is that despite the organisations identifying a reduction in costs through administration, paper, postage and time savings the cost savings benefits are at best estimated or indeed not stated due partly to the fact that reliable calculation and measurement models are not readily available.

Table 9.8 Identified benefits derived from the case studies

Benefits	*Projects 1-3	Case 4	Case 5	Case 6	Case 7	Case 8	Case 9	Case 10	Case 11
Central repository of information	X		X	X	X		X	X	X
Faster processing of data	X			X		X	X		X
Automation of data					X			X	X
Consistency of data	X			X	X	X	X		X
Less emphasis on human	X			X			X		
Reduced administration costs	X	X		X	X	X		X	X
Improved productivity	X			X		X		X	X
Structured workflows	X			X	X				
Unlimited users	X			X	X				X
No duplication of information		X	X	X		X	X	X	
More effective communication		X		X					X
Time savings	X	X		X	X	X	X		X
Less reliance upon paper	X		X			X			
Transactions electronic		X		X	X	X	X	X	X
Accountability	X		X	X					
Speed of processes improved	X	X	X	X	X	X	X	X	X
Speedier dissemination	X	X		X	X	X	X		X

* Project reviews 1-3 grouped together from a system point of view.
Note: The benefits above have been identified within the case studies under various headings. Other benefits within the case studies may exist although they have not been identified or indeed realised at this present time.

The penultimate chapter of this book has described in detail case study material received from a number or organisations operating within the construction environment. The range of implemented systems covers a wide range of managerial and operational areas including, customer service management, performance management, EDMS, supply chain RFID, facilities management, purchasing procedures, database building and digital pen and paper. The range or projects is perhaps more interesting and has produced some quite startling statistics on the levels of documentation that can be expected on major construction projects.

The chapter has also described in detail the various and numerous benefits that have been experienced from the implementation of the IT systems (these benefits have been summarised in Table 9.8). End-user passion for the implemented systems has suggested that the implementation of the IT systems was favourable and beneficial not only to the end users but also the organisation. However, within each of the reviews, we have also discussed and listed the lessons to be learnt: in essence 'the good things and the bad things' gleaned from the implementations. It will be these lessons learnt and a review of the Do's and Don'ts of IT implementation that will be the main focus of discussion within the last chapter.

CHAPTER 10

IT implementation: the do's and don'ts and the lessons learnt

This chapter will consider:

- The use of the PICA cycle.
- The Do's and Don'ts of implementation.
- Lessons learnt from the projects and cases.
- A summation.

10.1 PLAN, IMPLEMENT, CONTROL AND ADJUST (THE PICA CYCLE)

The underlying assumption throughout the early chapters of this book and the projects and case reviews is that the IT implemented has been, subject to some rigorous form of planning and development.

The planning and development being based on some suitable model which ensures that the fullest attention to detail has been given a suitable model to consider when implementing IT is the PICA cycle. We have taken the basic PICA cycle and expanded it to encompass a range of the detailed issues that require to be addressed when approaching the implementation of IT (Figure 10.1).

Where the work to be undertaken is considered in great detail i.e. people consciously consider a range of alternatives, they 'Plan'. This allows a range of positive alternatives to be considered and reduced to one plan which is chosen as the one to 'Implement'. When the plan is operationalised it is incumbent upon those charged with its success to ensure that they monitor what happens in action and take some form of corrective action as and when required to ensure that the planned outcomes are attained i.e. they exercise 'Control'. When all desired outcomes have been achieved, the operation of the plan provides a wealth of information which can be fed forward into the next area or activities for action i.e. 'Adjust' the next action.

As each element of the PICA cycle is considered (and a number of the sub-components) it will throw up a range of questions that have to be answered before the users can be being fully confident of moving on. The answers to these questions have led us to develop a range of Do's and Don'ts.

These Do's and Don'ts are not intended to be exhaustive, rather they should serve to make you think about the questions you have asked, or are about to ask of your IT solution. The case reviews and the projects have shown that the range of questions posed can be narrowed down and made manageable.

Starting with 'Notion' as the launch point we will move round the PICA cycle in a clockwise fashion to consider a number of the key components and resulting activities within the business. Not all will be considered in the same detail: given that the front end sets the scene for the rest of the action, the level of detail here will be greater than that given to the latter elements. As we consider each of the PICA elements we will suggest a range of Do's and Don'ts.

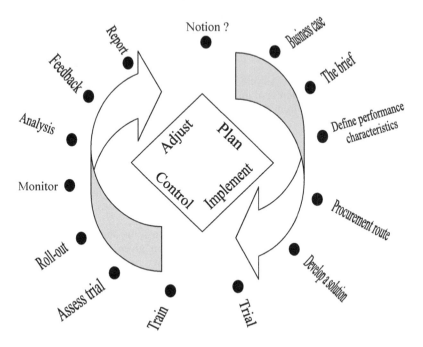

Figure 10.1 Planning for IT implementation

10.2 PLAN

When looking for a definition that best fitted what we intended to convey by 'Plan' we came to a very rapid conclusion that many of the definitions were too 'pedantic' in nature and would tend to drive the reader down the line of considering a true project management definition of 'Plan' i.e. with time constraints, resource considerations, cost implications and development of relationships with other activities. Thus we have decided that, in the PICA sense, 'Plan' can best be summed us by saying that we intend to achieve a desired state and in order to reach this state we must do specific 'things' in a distinct order, to bring about the desired outcome.

Having said what we see Plan as being, it then allows us to consider the various sub-elements i.e. the business case, the brief, choosing the procurement route, definition of the performance characteristics, and a lead-in to the Implementation phase.

Notion

The basic driving force behind any IT development or implementation work hangs on a notion generated by some person within the business. Whether this individual is within an IT department or unit is somewhat irrelevant: what is important is that a 'champion' of the notion is clearly discernible. They will have to ensure that it develops from an embryonic notion into a fully fledged, functioning, application or system that brings benefit to the business. There are clearly then a number of Do's and Don'ts associated with such a position:

- **DO** ensure that everyone is clear as to whom the notion's 'Champion' is. The clarity of a single reference point cannot be over emphasised. Accountability springs from this single reference point and helps to allay people's fears about who will manage the process.
- **DO** get support before you start development work: include all levels of the company. No matter where the notion for the solution originated, the bottom or the top, it must be fully supported by senior management (buy-in at this level is essential in order that the requisite resources will be committed). Supported solutions seem to be better received and more enthusiastically interacted with by other members of staff within the organisations.
- **DON'T** think that a notion developed in isolation and acted upon by a small number of individuals will have much chance of success.

Having ensured that the notion is robust and supported by an appropriate champion, the focus then moves on to defending it against other competing demands for the organisations valuable resources. This defence can be aided and made straightforward in a robust business case.

Robust business cases

A business case provides a broad range of information which enables management to make informed and rational decisions about whether a specific proposal should go ahead or otherwise. The business case provides an assessment of the value of investing in a specific proposal vis-à-vis other proposals in terms of potential benefits and the anticipated costs. Developing a business case is, by necessity, an iterative process producing further levels of detail at each iteration as the level of detail and understanding within the proposal unfolds. In its first guise it may appear as a one or two page document which outlines the notion at a strategic level: its main aim being at this point simply to gain approval to develop the business case further.

Well produced business cases have invariably been found to be team-efforts and represent a broad viewpoint from a range of key stakeholders. This consensus of understanding and commitment having been achieved early on acts as further support for the development of the proposal within the business case.

The business case helps to justify the IT investment by providing a framework which shows that investment should be made in the proposal as opposed to other areas of development e.g. machinery, vehicles, staff development, and so on. The business case also serves as a useful tool which forces the Champion and others to fully think through the implications of the proposed solution, including its impact on people. The business case also helps the identification and evaluation of the risks arising through not realising the benefits that are available, and planning which ensures the predicted return on the investment made is achieved.

Part and parcel of the process of actually developing a business case entails undertaking a range of tasks which include:

- Identifying the key stakeholders. A stakeholder being defined as any person or representative of an organisation who has a vested interest in the outcome of the IT implementation or whose opinion may be required to be accommodated.
- Identifying the options such that the development team do not concentrate on specific technologies, products or methods at the expense of other possibilities: all possible ways of meeting the demands of the successful solution should be examined and explored in detail.
- Appraise the 'do nothing' option - when assessing the costs associated with each option it is important to know the costs of the 'do nothing' option so that meaningful comparisons can be made.
- Map corporate and strategic alignment of the IT proposal; it should align with the organisation's corporate strategy (which hopefully is explicit).
- Vett the options; draw up a shortlist of the favourable options by considering the following questions:
    Does it indeed align with the stated corporate strategy?
    Is the proposed technology (both hardware and software) sufficiently robust and proven?
    Will the costs associated with the implementation be significantly outweighed by the resulting benefits?

The assessment of costs associated with implementing an IT solution includes not only the up front direct capital costs, but also a number of direct ongoing costs and a range of indirect costs which may be perceived as hidden costs. These three types of cost include e.g. direct capital costs associated with hardware and peripherals, packaged and customised/bespoke software, software development, systems integration and of course, training. The direct ongoing costs or 'out-of-pocket' expenses that occur over the lifecycle of the investment would include the likes of: maintenance contracts, licensing fees, salaries of internal IT support staff, ongoing training, telecommunications charges and, hardware or other insurances. The hidden or indirect costs are often forgotten. It should always be borne in mind that these costs (over time) can amount to much more than the direct costs. Examples of the indirect or hidden costs include: management time (vitally important if a full economic costing basis is used, since recovery of this cost will be sought form some revenue centre), staff time spent in training (before, during and after implementation), downtime due to device failure (from whatever cause e.g. replacement units if lost/broken, overload, etc.), an initial drop in productivity if the training has not been 100% effective, installation and updating bottlenecks and slow-downs and, corporate overheads which are often assumed to be simply 'absorbed'.

Having considered and discussed the sources and causes of cost in some detail, it is right that we consider the benefits in some detail also. Again, there are three areas associated with the benefits: those that are tangible, those that are intangible, and those associated with having 'perceived value' for the stakeholders.

The tangible benefits are normally relatively easy to quantify and, normally, result in direct forms of economic benefits e.g. reduced travel between operational units, a reduction in the number of administration staff and also on the time spent on routine administration itself, reduced reliance on, and utilisation of, paper. These benefits should be quantified by comparing the proposed solution with the 'do nothing' option – an exercise in business process mapping via tools such as 'Protos' will soon highlight where the gains are to be made. At the very least, the business needs to develop robust skills in information gathering. A discussion here on 'Performance Management' would be appropriate but falls out-with the focus of the book (for those interested visit sites such as www.g2bs.com).

The intangible benefits offer a slightly greater degree of complexity when seeking to gather sufficient data on them that will allow their incorporation within your business case. They play a key part in the decision making process and as such it is imperative that they be included. They can often be of significant value since they include areas that are important but difficult to actually pin-down e.g. increased end-user satisfaction, an increase in morale.

When it comes to establishing 'perceived value' to the various stakeholders, what has to be borne in mind is that the benefits that you identify will have a differing value for each of the interested stakeholders. Beauty is in the eye of the beholder, and as such will vary from stakeholder to stakeholder. Be attuned to this and aware of likely effects through the development of a matrix which allows the modelling and manipulation of benefits against stakeholders.

Finally, in terms of issues associated with costs, it is important to include at least a small section which details and deals with the 'risks' inherent in the implementation. Risk associated with the implementation comprises two main categories:

- Those risks associated with the implementation processes themselves and the desire to ensure that the solution 'suits or fits all'; and,
- The risk that the new, improved, solution system fails to deliver the benefits that had originally been envisaged and planned for.

There are a number of risk management tools available for use and a suitable approach for the organisation will be contingent upon a range of factors particular to the business itself. A risk matrix should be established which identifies the risks, assigns a (weighted) probability of the risk happening and, provides an indication of the effects of the risk occurring and how the risk may be mitigated.

The cost-benefit analysis can then be developed, now that we have collected all relevant information from the various sources. Cost-benefit analysis (CBA) is the systematic process of comparing the full range of costs associated with the implementation of the IT against the benefits that you see the organisation deriving from the implementation.

Once the analyses have been completed and a preferred option has been selected, the documentation of the business case can proceed, along with planning of how it will be presented.

As and when the business case documentation is completed, the task at hand is far from over: even the best analysis and documentation will be useless unless the decision makers 'buy-in' and give the necessary approvals. Here the role of the 'Champion' becomes crucial to the success, or otherwise, of the proposal. Assuming that the case is well supported and received, we can then move on to consider the development of the brief into a more substantial document and review its role in the overall implementation process. The business case clearly indicates there are a number of Do's and Don'ts to be aware of:

- **DO** find a champion and allow them to run with the implementation.
- **DO** develop a well rounded supporting case for the implementation.
- **DO** remember that costs and risks are important factors in decision making.
- **DON'T** rule out intangible or perceived benefits as being too 'obscure'.
- **DON'T** skimp on the business case: it's often the document that can win a difficult argument.

The Brief and its development
The Brief is to be viewed as a key document in its own right since it provides the foundations upon which the other documents can, and will be, built. It also helps in determining the direction and scope of the implementation project and acts as a formative mechanism in the contract between any project team/s selected and engaged and, the management of the business will rely on it for later monitoring and assessment of success. Any material change to the contents of the Brief will need to be referred to management for their action unless such decision making power and authority has been devolved to the implementation Champion. The brief fulfils a specific purpose in that it outlines the objectives and background/context for the overall implementation and also outlines the project management framework to be adopted for the delivery of the solution, including:

- how success of the various phases will be measured; and,
- what outputs to be delivered and when; and,
- the allocated budget and associated resources; and,
- the key (and detailed) project activities and milestones; and,
- the implementation governance and reporting requirements and arrangements; and,
- the approach to stakeholder and risk management; and,
- how assumptions, issues and constraints will be dealt with; and,
- the inter-relationship related other ongoing or planned projects; and,
- the relevant guidelines and/or standards to be used e.g. ISO 17799; and,
- level of reviews to be undertaken and quality assurance to be applied; and,
- how the lessons learnt will be captured and feedback into the business.

Some fundamental questions surrounding the brief should be addressed and answered prior to putting the IT implementation into action:

- Does the Brief accurately reflect the mandate for the IT implementation?

- Does the brief form a solid base from which to not only initiate a project, but to build a composite programme around?
- Does the brief indicate how the end-users (both internal and external) will assess the functionality and acceptability of the finished product(s)?

If these questions have been satisfactorily addressed, then the process can move on to consider what should be included in the actual 'brief package' that is to be issued to those expected to deliver the solution. The following list is intended to provide you with a flavour of what could be included; the final choice is of course down to the organisation, its operating environment and its desired outcomes from the composite implementation process:

- A description of the background to the problem and also the anticipated solution.
- Performance characteristics which explain what the implementation seeks to achieve. This section may contain:
    - Details on the project objectives, an outline synopsis of the project scope a discussion on the project deliverables and/or desired outcomes, any areas or issues that have been excluded or omitted, and a rationale of why, discussion on any 'perceived' constraints and, the nature and range of any interfaces anticipated.
- Business Case containing:
    - An evaluation of how this work supports the wider business.
    - The rationale behind the chosen solution and a description of the end-users' quality expectations.
    - Precise commentaries on how acceptance will be measured.
    - Any risks that have been identified.
    - Discussion on any relevant previous work.
- A draft Project Plan which contains reference to:
    - The aim of the work, the anticipated outcomes and key benefits expected and key dates/milestones.
- A detailed scope appraisal which:
    - Details what falls within the scope of the work and what does not fall within the scope of the work.
    - The range of objectives, deliverables and outputs and dates.
    - Project success measurement criteria.
    - Key assumptions made which affect the work.
    - Know constraints such as regulations or legislation that must be complied with, guarantees that may be required, and so on.
    - Any external factors that may impact on the work and how the work impacts on those external to the business (or business unit).
- Work areas and key activities:
    - Details on the key work areas, who is responsible for them and the priorities for each.
    - Any inter-and intra-dependencies known.
    - Any overlaps with other work being undertaken.

  o   A brief description of other organisational projects currently underway or planned.
- A discussion on the key risks associated with the work, and their management.
- An organogram depicting suitable structures with annotations illustrating governance and accountabilities.
- A briefing which describes resources and funding arrangements i.e. cash flows, payments, staffing (including contract arrangements i.e. full or part time) key skills required, and the use of consultants/sub-contractors.
- Discussion on the nature of the key stakeholders and their relationship to the business i.e. are they internal to the organisation or external e.g. Local Authorities, business partners, etc.
- An appraisal of how communications should be effected i.e. who, what, where, when and how.
- Clear statement on the quality levels expected with details on the quality management/assurance approach.
- A statement on the current position.

From the discussion above it is clear that there is a range of Do's and Don'ts associated with the brief:

- **DO** undertake thorough and detailed analyses of the processes which will change or be influenced by the IT: it doesn't matter if you don't end up implementing something new in the way of IT; the process of communicating with staff is valuable in its own right. If well done, the analyses will show areas where improvements can be made in the processes even before implementation of the new solution.
- **DO** ensure that you have full support from all involved and particularly from those who will be required to justify the resource commitment. This support should be tacitly given before you start the implementation. Remember that all organisations comprise the same two things, namely people and jobs. Engage those at the work-face who will be affected by the solution's implementation.
- **DO** remember that there has to be a justifiable business case for the solution. Analyse business needs first and then hardware and/or software capabilities. Cost benefit analysis can be undertaken to show that the solution will bring about real gains, both monetary and process based. Business needs should be carefully considered before the eventual hardware and/or software is chosen. Not all providers of solutions are equal: some providers can offer a greater range and scope of hardware and/or software than others.
- **DO** look for ways to get more, better quality information: engage with those who may hold the 'nugget of gold'. Many advantages are to be gained from properly trawling the back office systems and extracting the information that someone has lovingly saved for future use.
- **DO** make sure all stakeholders, no matter how obscure their relationship may appear, are involved early in the brief development, specification drafting and testing of any solution. There may be a larger variation of

user skills and attitudes than first envisaged: their input at focus group meetings and trial sessions may throw up considerable issues that were never anticipated at earlier stages.

- **DO** try to keep it as simple as possible. Think 'empathy' for the end-user. A solution which needs the least amount of training and helps users feel that they don't need to learn any new skills is a preferable solution. Empirical evidence suggests that people like responsive and welcoming solutions: those asked to use friendly solutions often accept them for the original problem and then move on to try them for other problem areas.
- **DO** remember that your organisation has internal and external customers: use information to enlighten external customers as well as the internal customers. Remember these external customers are another valuable source of reports on the applicability and success of a proposed solution.
- **DO** make full use of updated information (preferably on a daily basis). Trends determined from analysis of any changes found in daily or weekly information can help establish beneficial results, even if the information is not seen as critical.
- **DO** remember that quality expectations are influenced by all stakeholders. If these quality levels are attained then they rightly can be viewed as an immensely strong benefit.
- **DON'T** hide the success of the solution away in a dark cupboard. Many of your major suppliers and/or subcontractors may wish to make use of the solution and engage in bringing about improvements in their, and your, composite business processes.
- **DON'T** forget that the solution is a tool: avoid the temptation to sell the IT as a gimmick. Concentrate on the benefits gained from the new improved process.

Define performance characteristics

Before we can discuss the setting of performance requirements and how to improve performance, it's necessary to define what performance actually is. On the face of it, it sounds simple, but in reality it is a highly complex issue. In business, people often mean very different things when they begin to discuss 'performance'. There are several aspects of performance, each of which contributes to the overall performance of an IT solution or application.

Computers have an ever increasing capacity to carry out tasks with speeds which make the mere mortal blink with astonishment. When a computer, for whatever reason, doesn't perform a task quickly, users tend to be somewhat disappointed. Systems and software developers often see the terms 'speed' and 'performance' as being one and the same and therefore interchangeable. However, to understand the range and nature and difference in the types of problems that can be encountered, the different aspects of performance must be considered. There are a number of aspects that coalesce to bring about the 'performance' we require: with limited space, we are only able to discuss a few from the following list:

- Computer speed/performance.
- Memory characteristics.
- From cold to warm start-up time.

- Scalability.
- End-user perceived performance.

These factors form the foundations for a broader appreciation and understanding of performance. Whilst it is true to say that some aspects of performance are primarily applicable to the client-side and associated systems and some to the specialist server-side systems, it also equally applicable to say that some apply to both. Understanding how each factor can contribute to the performance characteristics of a system helps in the analysis of the performance of any particular application. If we consider each factor we see that computing performance is what most people think about first when discussing software performance. Computing performance considers characteristics such as:

- The number of instructions required to execute a specific statement (measured in millions of instruction per second – MIPS).
- How data could and should be structured for greatest efficiency?
- Which algorithms could and should be used?

Obviously, which algorithms are used and how they are implemented are key factors in the overall performance of the processor and associated software. The selection and use of algorithms and data structures turns out to be only part of the performance picture. Consideration must be given to factors beyond simply computing performance if the desire is to produce truly high-performing software.

Memory characteristics
The amount of memory needed to run the software can be of crucial importance to the overall performance of a system. All modern operating systems provide and utilise a virtual memory system whereby disk space is used in place of physical random access memory (RAM). However, even with increases in hard-disk operating speeds and the resulting lowering of access times, the fastest hard-disk is still slower than the slowest memory module. USB flash memories provide a close approximation to the desired speeds. It's not uncommon for beta-versions of applications to perform exceedingly well whilst under development but to perform very poorly once deployed. This fall-off in performance being traced back to the fact that the software developers typically have considerably more physical memory in their workstations than the average user has. Programs that run comfortably on a developer's machine might require significantly more memory than is available when installed on a user's machine.

It's also important to remember that the solution program probably won't be the only application running on a user's machine: most users typically keep more than three applications running at any one time. If the solution program consumed all of their memory resources then the users rightly would tend to be non-plussed with the solution.

- **DO** be aware that software being developed which is to be deployed on machines with limited RAM means that you will need to design software with the target machines' configuration in mind.

End-user perceived performance
It may be argued that perceived performance is the most important aspect of performance. Given that performance is in the eye of the beholder, then their view is the one that counts. Users seldom contemplate measuring performance in milliseconds: what is important is how fast the program actually 'feels', rather than how fast it really is i.e. it is subjective. There are a great many ways to improve how fast an application feels to the user without actually making any of the actual program code run any faster.

Simple things such as changing the standard mouse cursor to a timer cursor while the application is busy helps make the user feel the application is working for them. Telling the user that 'I am busy processing, please wait' helps them understand what is going on inside the program.

- **DO** put yourself in the end-user's shoes – empathise with them and ask what would you want and expect if you were the user? Think 'must have, should have and nice to have'.

Procurement route
How best to actually secure the products and services required to bring the solution to fruition is a path well trodden by many who have stumbled. What suits one organisation may not suit another; indeed there may be policies or legislation that guides which procurement route can or cannot be utilised. The Do's and Don'ts which follow are designed to and guide you through some of the important questions surrounding the choice of procurement route:

- **DO** consider the broadest possible range of procurement routes: the same route may not be applicable in all problem areas and for all proposed solutions.
- **DO** keep an open mind, large software vendors may have systems ready which they can lift off the shelf.
- **DO** seek and use expert advice when drafting contractual documents. The construction industry has a number of standard contract formats for construction works, but little that applies for procurement of other services e.g. IT.
- **DO** ensure that part of the procurement process includes checking the solution in the environment that it will be used in. Ask for all hardware/software and systems to be tested and commissioned.
- **DO** remember that the research you have undertaken on the business capabilities of the hardware and software is dated: your future proofing of needs must be considered.
- **DO** decide on where information is to be, and can be, stored. Carefully consider if the storage is to be on local devices or on the Internet (by a service provider or via the company's own server): this may depend on the connection stability, sensitivity and the security of the device and the amount of memory/length of time you need to store data for.
- **DO** ensure that all solution components are compatible with each other and existing elements/systems.

- **DO** look at best and worst scenarios for usage of the solution and ask for commentaries on these in the tender documents. It helps sort out those who have thought the whole process through.
- **DO** make sure all types of user are involved in testing of the solution as there may be a larger variation of user skills and attitudes than expected: their feedback is valuable.
- **DO** make sure the supplier is able to provide you with answers to crucial questions: What happens if the system crashes? How long will it take to cure large scale problems? Can you revert to previous systems for a short period of time?
- **DON'T** assume that you are unique and need new software and/or hardware. You may be able to get a completely off the shelf system for the hardware and the software that you require. This will be a tried and tested solution and probably cheaper.
- **DON'T** just implement an IT solution in isolation: any research on other works can improve the likelihood of success even before implementing the new solution.
- **DON'T** forget there may be a range of options on how to best use the hardware/software.
- **DON'T** frustrate the end-users. Ensure the solution operates at the speeds promised and that the clear benefits to the users are well understood.

Having planned and defined the essential performance characteristics of the solution it then naturally follows that we should consider how it will be developed and implemented.

10.3 IMPLEMENT

Develop a solution
- **DO** consult widely on the views and needs of the various end-users. Realise early on that some staff are more available than others, and more or less receptive to new ideas. The new ideas often take time to be fully understood and appreciated. Spend time with the doubters and run through the range of benefits to be expected from the solution: show them clearly how their working day will improve. Reinforce the ease of operation of the solution and how little will change in the way they do things.
- **DO** have a range of fallback options available ready to insert into the solution plan. Organise demonstrations of hardware and software options in order that people can begin to get a 'feel' for the solution and its impacts.
- **DO** realise that software and hardware development delays can, and do, arise just like delays in any real-life construction project. Promise results that can be delivered and be happy to err on the side of caution for those that cannot be guaranteed.

- **DO** check, and check again that everything melds together as it was intended to do. Compatibility issues need to be ironed out before the solution is exposed to the end-users: test all hardware/software and the systems that interface with the solution. Once is fine, twice is better and three times is getting towards being good.
- **DO** ensure that you carry out tests of the solution in the environment that it will actually be used in. Real-time use of IT in as real, or as near real, a construction environment as possible may throw up issues that haven't been thought of in the development stages.
- **DO** keep in mind that designing the solution to be modular allows it to be used by other parts of the business and builds in a degree of scalability that facilitates its use across the wider business spectrum. Ensuring flexibility becomes the key to its success: seldom has a solution been put in place that was perfect on day one.
- **DO** try to keep it simple: people who have been asked to use very friendly solutions often try to use them for other applications as well.
- **DON'T** underestimate the amount of time required for developing a possible solution. Time spent now, in ensuring the proposal selected is the best possible route forward will bring about real benefits at later stages.
- **DON'T** be tempted to continually change solutions. Hardware changes at such a rapid pace that the onus becomes quite often simply to stay ahead of the game: avoid such thoughts. Find a framework that will deliver the performance requirements and agree to stick with it for a period of time instead of continually changing to the most recent, updated, versions unless they can be shown to offer real business improvements or are updated as part and parcel of contractual arrangements.

Trial

- **DO** try to understand that staff are people and they have feelings like you do. Be attuned to the feelings of staff: if they are not used to using, or dislike, a particular application they may see it as being the safer option to revert to older applications or worse still, a pen and paper.
- **DO** review and where ever possible, improve the original work process. Doing this before using the IT solution will ease adoption.
- **DO** make it clear to staff that the trial is just that, not the final, polished solution. Staff need to be told they are trialling a solution and as such still under development with room for improvement based on their feedback. This engenders ownership of the final solution and also stops them expecting perfection in the trial.
- **DO** communicate, communicate and communicate. Keeping in constant touch with users allows their questions and fears to be addressed and makes sure that the system is being used the way in which it was intended. If it is not being used as intended, then the constant communication provides ample opportunity to find out why and address it.

- **DON'T** assume that one trial which goes well means that everything is perfect. The greater the number of trials the greater the number of unknowns will be identified and remedied.
- **DON'T** see the trial as a cost, it is a key step in the overall implementation process and provides a wealth of useful information that guides future activities.
- **DON'T** be afraid to tell people of the changes which have come about as a result of the trial. They are improvements to the system/application which means that the work to be undertaken by the staff, who will use it, should be even better.

Having trialled the solution we are able to gather significant information on its effects on the people and the business processes. This information provides a framework from which to develop the intended control mechanisms.

10.4 CONTROL

Assess the Trial
- **DO** be open and honest about the trial/s.
- **DO** communicate the outcome of the trial to all interested stakeholders. A series of events which outlines the findings and the improvements is useful.
- **DO** make it clear to staff that their part in the trial is appreciated.
- **DON'T** make promises based on the results of the trial which are not supported by the findings of the trial.
- **DON'T** ignore the feedback from the trial users, they have been responsive and you should reciprocate.

Train
- **DO** give as full and as relevant training as possible: it is not cost; at the very least it is direct engagement with the staff. Many of the minor issues can often be identified and resolved through training sessions. Remind staff that help is always on hand and reinforce how this help can be accessed.
- **DO** keep an open mind as to the changes which might result from the training and trialling of a solution. The unexpected changes can actually bring about significant improvements in the development of the solution.
- **DO** review the frequency of use of the solution, particularly when users are getting to grips with the basics.
- **DON'T** be frightened to carry out audits to see how well, or otherwise, the training is going. Ask if it is hitting the right targets: objectives set should be reviewed and adjusted if necessary. Offer help where it appears to be needed.
- **DON'T** rush at the training: it seems to be more effective and work better if it distributed over a planned period. Bulk sessions may be required but only if they are for very general materials. Remember that staff have to become familiar with the system and it does take time.

Assess Trial

- **DO** use lots of methods of assessment. People who have been asked to use very friendly solutions often try to use them for other applications as well, How the end-user reacts to the technology is important and has a significant effect on the future roll-out: find out their best liked and worst liked features.
- **DO** follow-up on the initial reactions and improve the solution wherever possible.
- **DO** engage as wide a participant body as possible in assessing the trial. The greater the number of voices heard now the easier the roll-out.
- **DON'T** assume that minor 'glitches' with the trial can be ignored. Staff will remember little things that irked them.

Roll-out

- **DO** plan the roll-out in fine detail and have back-up plans ready.
- **DO** make sure that those charged with rolling out the solution have all necessary knowledge, equipment and support information.
- **DO** engage with the end-users as the roll-out gets under way to ensure they are happy with the process. Make them feel cared for.
- **DO** ensure that the solution is being used the way it was intended to be. If it is not, find out why and address it. It may have nothing to do with the solution itself but more to do with other business environment factors.
- **DON'T** expect the roll-out to go without a hitch, plan ahead and anticipate what might happen and be able to address the issues as they arise.

Monitor

- **DO** devise a robust monitoring policy and ensure that everyone is familiar with how it will operate.
- **DO** advise people that the monitoring is underway and their input is appreciated.
- **DO** engage in regular feedback sessions from the data already gathered and tell people how this fits in with the larger feedback plan.
- **DON'T** allow the monitoring to proceed as if in automatic mode. There must clearly be seen to be active participation.

Whilst the monitoring of solutions in action is enjoyable in its own right we should not forget that a broad range, and significant quantity, of data is being made available, and hopefully appropriately gathered. The data is only useful if we subject it to some rigorous form of analysis that transforms it into meaningful and useful information.

10.5 ADJUST

Analysis

- **DO** consider how the data gathered will be stored and processed. Who will actually undertake the collection and processing – if automated, remember to fit the collection and processing times around the core activities.

- **DO** ensure you have a clear plan regarding dissemination of the processed data and tell staff how and when the findings from the analysis will be made available.
- **DO** remember that humans will have to try and understand what the findings are trying to convey: keep the language at the most appropriate level.
- **DON'T** shroud the analysis in secrecy: rumours are damaging, stop them before they have a chance to develop.

Feedback

- **DO** plan for a concerted effort to ensure that effective feedback is provided to all involved. Disseminate relevant (to each stakeholder) findings from the implementation: even your external stakeholders may find that they can learn from the discussions in the feedback.
- **DON'T** rush the feedback. Better for it to take a short period of time to be properly developed than rushed out in a state of disarray.

Report

- **DO** compile reports appropriate to the reader's knowledge and needs.
- **DO** make the reports succinct.
- **DON'T** make them over elaborate: showmanship may be seen as an attempt to mask unpalatable items.

10.6 THE LESSONS LEARNT

The solutions implemented were invariably customised: this customisation ranging from 50% to completely. The number of instances where an off-the shelf solution fitted with the business needs was very small. The suggestion being that the bespoke products more fully matched the needs of the business. There may however be a deeper underlying reason and that is to do with contractual relationships found within the industry and a prevailing lack of full appreciation of IT and what it can do for the business. This is supported by the fact that only 50% of the cases integrated the implemented solution with other IT. The solution can perhaps best be perceived off as an island within the sea of IT.

> **Action:** pin down the business needs.
> **Action**: ensure integration issues are considered and resolved.

The organisations involved in the implementations have been completely open and honest with us in providing data on their experience. The data provided has shown that a number of issues are clearly of importance to the organisations and yet, an important aspect that received scant attention was that off payback. Only one organisation was able to tell us what the return on investment had been, but that figure shows clearly that the benefits and savings have been far in excess of the initial outlay (a 200% return).

Clearly then there is a need for organisations to develop a return on investment model that kicks in as soon as the solution is being implemented. Only then can the fullest data be captured and processed to provide information that supports future implementations.

Action: develop a cost/benefit analysis model.
Action: put in pace a robust data gathering mechanism.

The timeframes reported were consistent in that the durations were of a year or less. Within the frameworks returned, each organisation was explicit in their clarity of stages and the need to keep timeframes tight. The fact that most allowed only a few months for the tender stage exemplifies the drive to 'keep it tight'. Keeping the implementation within a strictly controlled project environment works.

Action: develop as precise an implementation plan as possible.

In the majority of cases reported, the IT support for the implementation and the implemented solution was sourced externally. Coupled with this general external support is a significant amount of training (provided by both internal and external means). In one case, even though the system cost was relatively low, and the solution was an off-the shelf product, the company still gave over a significant amount of time to training. The training helping the staff to accept and utilise the solution being implemented.

Action: develop a training plan early on.

The benefits derived from the implementations were clearly demonstrable and can be grouped into three broad areas: those surrounding business process improvements, those associated with the data itself, and those influencing the humans working within the businesses.

The benefits to the business processes included refined workflows which brought about productivity gains. Indeed the greatest common benefit was the perceived increase in speed of the underlying processes. Also, the improved processes brought about perceived benefits in the communications effected within the businesses. The fact that transactions were committed electronically meant that there was clarity in the communication and accountability was seen as being more precise and so added to the general business process improvements. The speed by which information could be shared was flagged up by a significant number of the respondents as being a real plus as was the ability to include a greater range and type of individuals within the communication process.

Centralisation of otherwise disparate data and information repositories and a resulting increase in efficiencies surrounding access times and routines was seen as a real benefit. The management of the data and information circulating within the organisations was more efficient and less of a drain on the humans.

The reduction in the reliance on paper was seen by many as a significant improvement not just in terms of business processes but also in terms of the nature of the work activities undertaken. Many mundane activities were removed and so the quality of the work was felt to have improved. Even the removal of the reliance on human data input was seen in a number of organisations as being extremely worthwhile. The removal of duplication of effort, and multiple requests for the same data or information being seen as a result which pleased many.

The range and breadth of benefits gained by each of the organisations has shown that they all completed the implementation with a positive view of the processes actually involved, the additionally gained from the implementation itself and also the speed with which the implementation could be integrated within the underlying business processes and systems. The actual increases in the business processes themselves have been discussed previously and are worth reiterating as a real gain for the organisations.

10.7 FINAL COMMENTS

The book contains 10 Chapters, each of which we hope have added to the general body of knowledge that seeks to discuss and expand how IT is best implemented. Within Chapter 1 we set out a broad landscape of a range of elements within the general concept of IT and how it relates to the way construction is delivered in today's hectic business environment. Chapter 2 discussed the operation of the construction industry and the range of organisations that were likely to be encountered and indeed engage in IT activities, Chapter 3 focused on the role humans have to play in all things IT centric within the organisation and also the broader industry. The role humans play on IT initiatives was also considered along with the effects of IT on humans. Chapter 4 concentrated on the actual business processes executed on a day to day basis and how they are affected and improved by appropriate application of IT. Chapter 5 delved into the knowledge repositories found within the organisations and the broader industry, and how we manage this widely differing range and breadth of knowledge bases. Chapter 6 delved into the issue of knowledge much further by considering how we captured and processed information at a range of organisational levels. Chapter 7 reviewed a range of applications (tools) found within the industry, with some detailed discussion on a few of the common ones. Chapter 8 looked at IT implementations which had gone before and the mistakes often made during these implementations. Chapter 9 reviewed a range of participating organisation's efforts at implementing IT and the resulting impacts. Chapter 10 provides a consolidation of the positive points taken from the cases reviewed and provided a range of suggestions as to what to do or not to do, as the case may be.

We said right at the commencement of the book that it was not intended to be the panacea to all IT issues. We openly admitted to not being IT experts but saw this as being a factor in our favour. We have to thank the companies who graciously provided us with free access to some very sensitive data and made us feel welcome at every juncture. We hope you will agree that the space has been used wisely to discuss the details we felt were important and that you have enjoyed reading this book.

REFERENCES

AELR, Walker v Northumberland County Council [1995] 1 All ER 737.

Ahmad, IU., Russell, JS. and Abou-Zeid, A., 1995, Information technology and integration in the construction industry. *Journal of Construction Management and Economics*, **13**, pp. 163-171.

Alkass, S., Mazerolle, M., Tribaldos, E. and Harris, F., 1995, Computer aided construction delay analysis and claims preparation. *Journal of Construction Management and Economics*, **13**, pp. 335-352.

Anumba, CJ., Aziz, Z. and Obonyo, EA., 2003, Mobile Communications in Construction – Trends and Prospects, *Innovative Developments in Architecture, Construction and Engineering, Loughborough University*, pp. 159-168.

Anumba, CJ., Egbu, C. and Caririllo, P., 2005, Knowledge management in construction, Blackwell Publishing, Oxford.

Aranda-Mena, G. and Stewart, P. 2005, Barriers to e-business adoption in construction international literature review, *COBRA RICS Annual Conference, Queensland University of Technology*. 4th – 8th July 2005 (full document, electronic proceedings).

Aziz, ZUH. and Tah, JHM. 2002, Bluetooth Wireless Technology and Its Potential Applications for Construction Site Processes, *International Conference on Decision Making in Urban and Civil Engineering*, London.

Backbolm, M., Ruohtula, A. and Bjork, B., 2003, Use of document management systems – a case study of the Finnish construction industry. *Electronic Journal of Information Technology in Construction*, **8**, pp. 367-380.

Best Practice Group PLC. 2004, The little book of IT project mistakes, available at http://www.bestpracticegroup.com/10mistakes.htm.

Boehmler, R., 1998, Fully digital survey systems for property. *Journal of Structural Survey*, **16** No. 2, pp. 61-66.

Bollinger, AS. and Smith, RD. 2001, Managing organizational knowledge as a strategic asset. *Journal of Knowledge Management*, **5** (1), pp. 8-18.

Bolton, JE. Committee, 1971, Small Firms: Report of the Commission of Inquiry, Cmnd 4811, HMSO (London).

Bowden, S., 2001, Construction Site Information Needs, Arup. London, http://www.cite.org.uk/Resources_Case_Studies_IT_On_A_Construction_Site.html.

Bowden, S., 2002, Mobile Technology Construction Software, Arup, London, http://www.cite.org.uk/Resources_Case_Studies_IT_On_A_Construction_Site.html might need access date.

Bowden, S., 2005, Application of mobile IT in construction, PhD thesis, http://www.lboro.ac.uk/cice/Theses/Sarah%20Bowden%20final%20version.pdf

Bowden, S., Thorpe, A. and Baldwin, A., 2002, Usability Testing of Handheld Computing on a Construction Site.

Breetzke, K. and Hawkins, M. 2003, An introduction to project extranets and e-procurement, Royal Institute of Chartered Surveyors, available at, http://www.rics.org/Builtenvironment/Constructionprocurementandtendering/An+introduction+to+project+extranets+and+e-procurement.htm.

Cheng, EWL., Li, H., Love, PED. and Irani, Z., 2001, Network communication in the construction industry, *International Journal of Corporate Communications*, **6** No. 2, pp. 61-70.

CITB, 2002, Managing Profitable Construction: The Skills Profile, CITB, July, London.

Crossan, M. and Inkpen, A. 1995, The subtle art of learning through alliances. *Business Quarterly*, **60**, pp. 68-78.

De la Garza, JM. and Howitt, I. 1998, Wireless communication and computing at the construction jobsite. *Automation in Construction*, **7**, pp. 327-347.

Department of the Environment. 1995, Bridging the gap: construct IT. Salford: Salford University.

Disterer, G., 2002, Management of project knowledge and experiences, *Journal of Knowledge Management*, **6**(5), pp. 512-520.

Dixon, NM., 1999, The Organizational Learning Cycle: How we can learn collectively. Gower Publishing Limited, Aldershot.

Doherty, P., 2000, e-Business and the Changing Face of the Construction Industry, the Digit Group.

EC 96 Council Decision 97/15/CE of 9.12.1996, OJ L 6 of 10.1.1997, p. 25.

Egan, J., 1998, The Egan report – Rethinking Construction. Report of the construction industry taskforce to the Deputy Prime Minister.

El-Ghandour, W. and Al-Hussein, M. 2004, Survey of information technology applications in construction, *Journal of Construction Innovation*, **3**, pp. 83-98.

Elvin, G., 2003, Tablet and Wearable Computers for Integrated Design and Construction, *American Society of Civil Engineers Construction Research Congress, March 19th, Hawaii.*

Fahey, L., Srivastava, R., Sharon, JS. and Smith, DE., 2001, Linking e-business and operating processes: The Role of Knoledge management. *IBM Systems Journal*, **40** (4).

Fenn, P., Lowe, D. and Speck, C., 1997, Conflict and dispute in construction, *Construction Management and Economics*, **15**, No. 6, pp. 513-518.

Fernie, S., Green, SD., Weller, SJ. and Newcombe, R., 2003, Knowledge sharing: context, confusion and controversy. *International Journal of Project Management*, **21**, pp. 177-187.

Fidelman, U., 2002, Temporal and simultaneous processing in the brain: a possible cellular basis of Cognition, Kybernetes 31, 3/4 pp. 432-481.

Florén, J., 2002, Measuring the cost and value of Anoto functionality implementations compared to PDA implementations, Anoto Functionality Publications, Sweden.

Fong, PSW., 2003, Knowledge creation in multidisciplinary project teams: an empirical study of the processes and their dynamic interrelationships. *International Journal of Project Management*, **21**, pp. 479-486.

French, JRP. and Caplan, RD. 1973, Organisational stress and individual strain, in (The Failure of Success, Morrow, Ed.).

Fryer, B., 1998, The practice of construction management, 3rd Ed, Blackwell Science.

Gidado, KI. and Nichols, M. 2002, The Current State of Use of Electronic Document Management System by UK Architectural Practices. *In electronic proceedings of the 2nd International Conference on Decision Making in Urban and Civil Engineering,* London.

Griffith, A. and Headley, JD. 1998, Management of small building works, *Construction Management & Economics,* **16**(6), pp. 703-709.

Gyampoh-Vidogah, R., 2000, Guidelines for implementing EDMS in the construction industry, University of Wolverhampton, document available at: http://www.itconstructionforum.org.uk/publications/publication.asp?id=819.

Hammant, J., 1995, Information Technology Trends in Logistics, *Journal of Logistics Information Management,* **8**(6), pp. 32-37.

Hoch, SJ. and Deighton, J. 1989, Managing What Consumers Learn from Experience, Journal of Marketing, **53**, pp. 1-20.

Homan, G., Hicks-Clarke, D. and Wilson, A., 2001, The management development needs of owner managers/managers in SME's: mapping the skills and techniques, The Manchester Metropolitan University, Manchester.

Inkpen, A., 1998, Learning, knowledge acquisition, and strategic alliances. *European Management Journal,* **16**, pp. 223-229.

Institution of Civil Engineers (ICE), 2002, Mobile Communications and Hand Held Devices Field and Asset Management Use Briefing Sheet.

ISO 9000, 2005, Quality management systems, British Standards Institute, London.

ISO 17799, 2005, Code of Practice for Information Security Management, British Standards Institute, London.

ISO 15489, 2001, Information and documentation-records management, British Standards Institute, London.

Juran, JM. and Gryna, FM. 1993, Quality Planning and Analysis, 3rd Ed, McGraw-Hill, London.

Kangari, R., 1995, Construction claim documentation in arbitration. *Journal of Construction Engineering and Management,* 121, pp. 201-208.

Kasvi, JJJ., Vartiainen, M., and Hailikari, M., 2003, Managing knowledge and knowledge competences in projects and project organisations. *International Journal of Project Management,* **21**, pp. 571-582.

Kumaraswamy, MM., 1997, Conflicts, claims and disputes in construction, *Journal of Engineering, Construction and Architectural Management,* **4**, No. 2, pp. 95-111.

Latham, M., 1994, Constructing the team. Final report of the Government/industry review of procurement and contractual arrangements in the UK construction industry. HMSO, London.

Lingard, H. and Holmes, N. 2001, Understandings of occupational health and safety risk control in small business construction firms: barriers to implementing technological controls, *Construction Management & Economics,* **19**, No. 2, pp. 217-226.

Love, PED., Tse, RYC., Holt, GD. and Proverbs, DG., 2002, Transaction costs, learning & alliances, *Journal of Construction Research,* **3**, pp. 193-207.

Marsh, L. and Flanagan, R. 2000, Measuring the benefits of information technology in construction, *Journal of Engineering Construction and Architectural Management,* **7**, No. 4, pp. 423-435.

Meticulus Solutions Limited, 2002, New Product Launch: meticulive throws out the paper archive, http://www.meticulus.com/MeticulistDefault.htm.

Mohamed, S., 2003, Web-based Technology in Support of Construction Supply Chain Networks, *Journal of Work Study,* **52**(1), pp. 13-19.

Molad, C. and Back, E. 1995, EDI's role as an enabler for electronic commerce and information integration, *Journal of Engineering Construction and Architectural Management,* **2**, No. 2, pp. 93-104.

Motwani, J., Madan, M. and Gunasekaran, A., 2000, Information Technology in Managing Global Supply Chains, *Journal of Logistics Information Management,* **13**(5), pp. 320-327.

Mzenda, V., 2002, Intranet and extranet – the benefits, a short review, available at http://www.kizukigroup.com.

National Archives of Australia, 2005, 'Glossary', available at www.naa.gov.au/recordkeeping/er/guidelines/14-glossary.html.

National Statistics Office (NSO) 2002, Annual Abstract of Statistics, HMSO, London.

National Statistics Office (NSO) 2005, Annual Abstract of Statistics, HMSO, London.

Newton, P., 1998, 'Diffusion of IT in the Building and Construction Industry'. CSIRO, *Building for Growth Innovation Forum, Sydney* 4-5 May 1998.

Nonaka, I. and Takeuchi, H. 1995, The knowledge creating company: how Japanese companies create the dynamics of Innovation. Oxford University press, Oxford.

O'Brien, M. and Al-Soufi, A. 1993, Electronic data interchange and the structure of the UK construction industry, *Journal of Construction Management and Economics,* **11**, pp. 443-453.

Ribeiro, FL. and Lopes, SJ. 2001, Knowledge-based e-business in the construction supply chain. *In: Proceedings of COBRA 2001, The Construction and Building Research Conference of the RICS Research Foundation, 3rd – 5th Sept 2001, Glasgow, Glasgow Caledonian University,* pp. 711-720.

Royal Institute of Chartered Surveyors (RICS), 2003, Electronic Document Storage Legal Admissibility, 2nd Ed, RICS Books.

Raynes, M., 2002, Document Management: is the time right now?, *International Journal of Work Study,* **51**, No. 6, pp. 303–308.

Robertson, HW. and Sommerville, J. 1994, Total Quality Management in The Morrison Construction Group: Managing Transition and Change *(Proceedings of Eureka Conference, Hamar/Lillehammer),* pp. 326-331.

SFEDI (2002) Small businesses skills assessment, Small firms Enterprise Development Initiative, Sheffield.

Sommerville, J. and Craig, N. 2002, Sysdox! As a Decision Support Tool, *International Conference on Decision Making in Urban and Civil Engineering,* London 6th–8th December 2002. (Electronic proceedings).

Sommerville, J. and Craig, N. 2002, Systems Thinking in Construction Management! A case review, 2nd International Conference on Systems Thinking in Management, Salford University. 22nd–24th April 2002, session A3, pp. 8–13.

Sommerville, J. and Craig, N. 2003, Cost Savings of Electronic Document Management Systems: The Hard Facts, COBRA RICS Annual Conference, University of Wolverhampton. 1st–2nd September 2003, Vol 1, pp. 279–287.

Sommerville, J. and Craig, N. 2004, Information Processing Using a Digital Pen and Paper, International Conference on Construction Information Technology (INCITE 2004): Managing Projects through Innovation & IT Solutions, Langkawi, Malaysia: 18th–21st February 2004, Vol 1. pp. 217–224.

Sommerville, J., Craig, N. and McCarney, M., 2004, Document Transfer between Distinct Construction Professionals. *COBRA RICS Annual Conference, Leeds Metropolitan University,* 7th–8th September 2004, **1**, p. 136 (full document, electronic proceedings).

Sommerville, J., Craig, N. and Ohlstenius, O., 2004, Handwriting Recognition: Improving the Effectiveness of Construction Project Information, 4th International Post Graduate Research Conference (IPRC 2004): Salford University. 1st–2nd April 2004, Vol 1, pp. 274–283.

Sommerville, J. and Craig, N. 2005, The digital revolution with pen and paper, CIB W92 Procurement symposium, Arizona State University, 7th–10th February 2005, Vol 1, pp. 315-323.

Stephenson, P. and Turner, P. 2003, Electronic Document Management Systems in Construction: A Project Based Case Study Implementation. *Proceedings of the Architecture, Construction and Engineering conference,* Loughborough University 25-27th June 2003.

Stewart, RA. and Mohamed, S. 2003, Evaluating the value IT adds to the process of project information management in construction, *Journal of Automation in Construction,* **12**, pp. 407-417.

Stewart, RA., Mohamed, S. and Marosszeky, M., 2004, An empirical investigation into the link between information technology implementation barriers and coping strategies in the Australian construction industry, *Journal of Construction Innovation,* **4**, pp. 155-171.

Tam, CM., 1999, Use of the Internet to enhance construction communication: Total Information Transfer System, International, *Journal of Project Management,* **17**, No. 2, pp. 107-111.

Then, DSS., 1995, Computer aided building condition surveys, *Journal of Facilities,* **13**, No. 7, pp. 23–27.

Thomson, A. and Gray, C. 1999, Determinants of management development in small businesses, Journal of Small Business and Enterprise Development, **6**, No. 2 pp. 113-127.

Vidogah, W. and Ndekugri, I. 1998, A review of the role of information technology in construction claims management, *Journal of Computers in Industry,* **35**, pp. 77-85.

Wantanakorn, D., Mawdesley, J. and Askew, H., 1999, Management errors in construction, *Journal of Engineering, Construction and Architectural Management,* **6**, No. 2, pp. 112-120.

Weippert, A., Kajewski, SL. and Tilley, PA., 2003, The implementation of online information and communication technology (ICT) on remote construction projects, *Journal of Logistics Information Management,* **16**, No. 5, pp. 327-340.

Wen, HJ., Yen, DC. and Lin, B., 1998, Intranet document management systems, *Journal of Intranet Research,* **8**, No. 4, pp. 338–346.

Wyer, P., Mason, J. and Theodorakopoulos, N., 2000, Small business development and the learning organisation, *International Journal of Entrepreneurial Behaviour and Research,* **6**, No. 4, pp. 239-259.

Xerox, 2002, Convergent Document Technology an IT manager's guide, http://www.istart.co.nz/index/DOCC199/F14907.

Yeomans, SG., Bouchlaghem, NM. and El-Hamalawi, A., 2006, An evaluation of current collaborative prototyping practices within the AEC industry, *Automation in Construction*, 15, pp. 139–149.

http://www.dti.gov.uk/SME4/define.htm

Index

eBooks

eBooks – at www.eBookstore.tandf.co.uk

A library at your fingertips!

eBooks are electronic versions of printed books. You can store them on your PC/laptop or browse them online.

They have advantages for anyone needing rapid access to a wide variety of published, copyright information.

eBooks can help your research by enabling you to bookmark chapters, annotate text and use instant searches to find specific words or phrases. Several eBook files would fit on even a small laptop or PDA.

NEW: Save money by eSubscribing: cheap, online access to any eBook for as long as you need it.

Annual subscription packages

We now offer special low-cost bulk subscriptions to packages of eBooks in certain subject areas. These are available to libraries or to individuals.

For more information please contact webmaster.ebooks@tandf.co.uk

We're continually developing the eBook concept, so keep up to date by visiting the website.

www.eBookstore.tandf.co.uk

.